Conquering Statistics

Numbers without the Crunch

Conquering Statistics

Numbers without the Crunch

JEFFERSON HANE WEAVER

PLENUM TRADE • NEW YORK AND LONDON

Library of Congress Cataloging-in-Publication Data

Weaver, Jefferson Hane.
    Conquering statistics : numbers without the crunch / Jefferson
Hane Weaver.
        p.   cm.
    Includes bibliographical references (p.   -   ) and index.
    ISBN 0-306-45572-2
    1. Statistics.   I. Title.
QA276.12.W374   1997
519.5--dc21   96-39660
                                                              CIP

ISBN 0-306-45572-2

© 1997 Jefferson Hane Weaver
Plenum Press is a Division of Plenum Publishing Corporation
233 Spring Street, New York, N.Y. 10013-1578
http://www.plenum.com

10 9 8 7 6 5 4 3 2 1

Printed in the United States of America

This book is fondly dedicated to my
grandmother, Anna L. Wood,
with great love and respect.

Contents

Statistics and Serendipity in a London Haberdashery

Let nature and let art do what they please,
When all is done, life's an incurable disease.
—ABRAHAM COWLEY

Introduction

How would you like to have a book that would enable you to predict the winning numbers in a state lottery? How would you like to have a book that would enable you to win the jackpot at a slot machine in a Las Vegas casino whenever you decided you needed some extra cash? How would you like to have a book that would invariably guide you to picking the winning team when making a bet on sports with the neighborhood bookie? Well, we would like to have that very same book, so please send it to us in care of the publisher if you should happen to run across it.

Because our readers are much more interested in learning about deep intellectual topics, we have decided not to waste our time passing on foolproof schemes for striking it rich in the nation's casinos. Mind you, we

1

could go for the cheap applause by providing the winning numbers for all of the state lotteries that will be held next year. But passing out these numbers would certainly have no appeal to our readers, who are very intelligent people with much better things to do with their time than stuff their bank accounts with a few million dollars.

This book is about statistics. Now, statistics is not a popular subject for discussion in trade publishing circles because it is viewed with as much enthusiasm as the Ebola virus. This is not to say that statistics books are any more tedious than the textbooks for any other class one might take in college. But statistics books have very few pictures of fuzzy animals or attractive celebrities. Indeed, you would probably have a very difficult time naming a movie actress who could truthfully talk about the positive glow she felt after reading a few pages from her introductory statistics book every morning. Moreover, statistics has never really found its way into the popular culture like certain other branches of modern science. Many would-be pharmacists, for example, have not let their lack of formal education or moral scruples stand in their way of making a living by peddling narcotics on the street. Their desire to provide a quality product to the nation's addicts serves as a sad example for the rest of us of the depths to which money-grubbing individuals can sink.

Similarly, chemistry has become all the rage among disaffected loners and paramilitary groups who have learned of the wondrous variety of explosive devices that can be made with a few simple ingredients found in the kitchen cabinet. No one can doubt the sparkle in a young child's eye when he makes his first pipe bomb or packs his wagon full of explosives and leaves it in front of the neighborhood bully's house. Statistics admittedly suffers somewhat by comparison because you cannot really use statistics to get involved in the drug trade or to build devices of mass destruction. Moreover, it is difficult to highlight the principles of statistics in a television situation comedy because such basic concepts as bell curves and degrees of confidence do not readily lend themselves to knee-slapping jokes.

So statistics has something of an image problem. It is not considered to be a particularly glamorous topic for discussion. Most young boys, for example, would rather lead a football team to a national championship than memorize detailed tables of binomial probabilities. This inexplicable preference has been borne out in survey after survey. Sadly, statistics has no greater appeal among young girls, who would prefer to model designer gowns on a Paris runway or even date deviants instead of calculating the standard deviations of a given population. This reluctance of the nation's youth to embrace statistics is disheartening to the many dozens of individuals who have worked tirelessly to convince the rest of us that statistics is full of fun and excitement. Because most people do not believe that statistics is full of fun and excitement, however, these efforts have been largely ignored.

The real tragedy of this situation is that statistics, like broccoli, is a very good thing even though both may have a slightly malodorous aftertaste. Statistics is even more useful than broccoli in helping companies make decisions about the types of products they should produce, the ways in which they should market those products, and the production processes that are to be used to make those products. Broccoli, by contrast, is of very little use in formulating a marketing plan or operating a factory, even though it is a good source of iron.

Thus, statistics is an orphan in the national consciousness even though we have to deal with it almost every day. Every time we see a poll in which people reveal, for example, that they would rather wear purple lingerie to a wine-tasting party than to a funeral, we are, in a sense, using our knowledge of statistical principles. One of the most common elements of any such survey is the clarification usually noted during the survey regarding the possible margin of error (e.g., "plus or minus 3%"). This margin of error merely refers to the potential variations that could cloud the results of the survey. Moreover, statistics is used in almost every industry to monitor such things as the rate of defects among various products and, hence, the efficiency of the manufacturing process itself.

Of course such commercial applications would not necessarily impress those individuals who live in caves and eat raw animals because of their disdain for modern technological devices such as matches. But if these very same individuals like to gamble and wager their hard-earned piles of rocks and kindling in high-stakes poker games, then they may reconsider their negative attitudes toward statistics. After all, poker is ultimately concerned with probabilities, and statistics is the study of probabilities. Therefore, the habitual gambler and the statistician can find common ground with each other and perhaps bet their car titles together at the poker table. If they know how to calculate the probability that a particular card will be drawn, they can make a somewhat educated guess as to the most likely cards that will be dealt. Of course it is not always so easy to count the cards and thereby determine which cards are more or less likely to turn up.

The thought of such shared devotion to the pursuit of knowledge for the sake of knowledge is refreshing in an age as materialistic as the one we live in. But it also illustrates in a limited way the wide sweep of statistics and the many ways in which it can be used during both work and play. Many of the examples used in this book will further drive this point home because we will see a wide variety of situations and circumstances in which statistics can be brought to bear to solve a nagging problem.

Who Is Mr. Statistics?

The birth of statistics occurred in a tiny haberdashery along one of London's cobblestoned streets in the mid-17th century. Its discoverer, John Graunt, was a native of London who boasted no extraordinary intellectual talents but could claim an appreciation for numbers and sound record-keeping. Unlike most of his colleagues, however, Graunt also developed interests outside of the shop, particularly in the areas of mathematics and politics, that would later serve him in his investigations of the births and deaths of his fellow citizens.

The eldest of Henry and Mary Graunt's seven children, John, was born in 1620 in Birchin Lane, London, "at the Sign of the Seven Stars," where his father ran a draper shop. The family also lived at the same address, and Mary Graunt raised a noisy but happy brood of youngsters. John may have learned the tenacity that would later serve him so well in the Hobbesian state of nature that must have existed at the family table when all of the family members sat down for a meal. John rarely ventured beyond his London neighborhood. He received little formal education and was apprenticed to a merchant of small wares at a young age. While working in his master's store, Graunt learned how to manage a shop and, more importantly, how to succeed in the volatile London retailing business. In 1641, he married the former Mary Scott, with whom he had one daughter and three sons.

Graunt was soon the proprietor of his own store and worked hard to attract a loyal clientele. Despite the demands of his daily schedule, he usually spent several hours before sunrise reading and recording his thoughts on subjects dear to his heart. He also became a close friend of Sir William Petty, the author of a widely circulated book on "political arithmetic." Graunt's conversations and correspondences with Petty caused him to wonder about the dynamics of human society as a whole and, more specifically, about the mortality patterns of the human race. Graunt's curiosity somehow prompted him to begin reviewing a weekly church publication issued by the local parish clerks that listed the numbers of births, christenings, and deaths in each parish. These so-called *Bills of Mortality* also listed the causes of death, thus providing Graunt with a massive but unorganized mass of information about the ongoing drama of birth and death occurring all around him.

Graunt's investigations of the births and deaths of London parishioners did not arouse great enthusiasm among his friends or family. Indeed, his daily habit of plowing through the *Bills* was viewed by most of his friends as nothing more than an eccentric quirk or, at the very least, a morbid curiosity. Graunt himself recognized that nearly all of the persons who took copies of the *Bills of Mortality* made little use of them other than to "look at the foot, how the burials increased or decreased; and, among the casualties, what had happened rare, and extraordinary in the week current: so that they might take the same as a text to talk upon, in the next company; and withall, in the plague-time, how the sickness increased, or decreased, so that the rich might judge of the necessity of their removal, and tradesmen might conjecture what doings they were like to have in their respective dealings."[1] In other words, the *Bills of Mortality* were something akin to tabloid fare; they provided those hardy souls who managed to survive all of the scourges and diseases of 17th-century London with a grim satisfaction that they had managed to avoid the icy reach of death—at least temporarily. Of course the comparatively short average life spans enjoyed by the persons of that era guaranteed that any such feelings of relief would be short lived.

But Graunt was motivated by a nagging feeling that there was something of value to be gained from perusing dozens, and later, hundreds of these publications even though he did not have a clear idea at the outset of what he was attempting to find. From the beginning, however, Graunt took care to organize his data in a way that was probably inspired by his techniques for tracking his shop inventory: He devised tables that were not only easy to read but could also be easily updated. He was not concerned with developing any grand theories of the kind that Isaac Newton would later use to explain the motions of the planets and stars; Graunt merely wanted to find any underlying principles or themes amidst the thousands of yellowed pages of the *Bills*. Graunt recognized his work would be far more accessible and useful if he and any other future investigators could avoid having to review the dusty piles of documents cluttering his study. So he took great pains to "[reduce] several great confused Volumes into a few perspicuous Tables, and [abridge] such Observations as naturally flowed from them, into a few succinct Paragraphs, without any long Series of multiloquious Deductions . . ."[2]

Graunt soon discovered that the process by which the *Bills of Mortality* were compiled was not a clerking procedure that would normally be found in a shop or school. This procedure began when someone happened to find a corpse in a house or out on the street. A bell would be tolled and a person known as a searcher would go to the corpse and perform both a visual inspection and, in some cases, a primitive autopsy. Through these investigations, the searcher would determine—based upon his or her own extensive medical training or best hunch—the disease or casualty that had caused the death. Armed with this information, the searcher would file the report with the parish clerk. Once a week, the parish clerk would carry a list of all the burials and christenings that had occurred that week to the clerk of the hall. Acting as a master compiler, the clerk of the hall would compose a general account of all births, deaths, and christenings and oversee the printing and publishing of this account. Once published, copies would be sold to those families willing to pay four shillings per year.

Graunt's research revealed a vast panoply of perils stalking the residents of London. In the year 1632, for example, the most popular causes of death and the associated number of deaths were "chrisomes, and infants" (2268), "consumption" (1797), "fever" (1108), "aged" (628), "flocks, and small pox" (531), "teeth" (470), "abortive, and stillborn" (445), "bloody flux, scowring, and flux" (348), "dropsie, and swelling" (267), and "convulsion" (241). That London's medical community was second to none was shown by the diagnoses of less well-known maladies such as "rising of the lights" (98), "suddenly" (62), "king's evil" (38), "purples and spotted fever" (38), "drowned" (34), "worms" (27), "executed, and prest to death" (18), "grief" (11), "scurvey and itch" (9), "burst and rapture" (9), "falling sickness" (7), "dead in the street and starved" (6), "lethargie" (2), and "piles" (1).

That the clerk of the hall was able to keep up with the body count, given the relentless determination of the citizenry to drop dead in all variety of ways, was admirable. But Graunt soon found something even more interesting when reading between the lines of the yellowed *Bills*: He could use the tables he had prepared to make certain deductions about the population as a whole. Graunt's genius lay in his ability to realize that these deductions could be made only by comparing the changes in these mortality tables over many years. Only by such long-term (longitudinal) comparisons could the subtle, yet primordial, demographic changes affecting the population of London as a whole be ascertained.

Graunt's inventory of the diseases (real and imagined) that were preventing most Londoners from living beyond the ripe old age of 40 enabled him to make some generalizations that were to be of great use to those few officials who were concerned with the public health. Nearly one-third of all the deaths, for example, were caused by such things as thrush, convulsions, rickets, teeth, and worms; these deaths occurred primarily among young children and infants. Similarly, young children were the predominant victims of small-pox, swine-pox, measles, and bubonic plague. Graunt suggested that many of the deaths among the adult population could ultimately be attributed to the lack of cleanliness in the city itself as well as the quality of food and water available to most Londoners. Most city-dwellers of that time were not greatly concerned about dumping garbage and sewage in the streets or drawing water from contaminated wells and rivers. As a shop-keeper who was constantly reminded of the debris through which his customers trudged, Graunt probably had little difficulty recognizing the role played by inadequate sanitation in the mortality rates of the citizenry. But it would be too great a stretch to credit Graunt's work with having anticipated the discovery of any theory regarding the origination and transmission of disease through germs. Another benefit of Graunt's work, however, was that it showed in relative terms which diseases were actually causing the most deaths. As such, it revealed a comparatively low incidence of the bubonic plague, even though the very mention of the word "plague" still struck fear in the hearts of Englishmen, whose ancestors had seen nearly a third of their island's population wiped out by that disease nearly 300 years before.

Although Graunt's mathematical techniques were not terribly sophisticated, his analysis was quite impressive. He discovered that more boys than girls were born, women tended to live longer than men (even in those days), comparatively few people died from starvation, and deaths from disease did not vary greatly from year to year except when an epidemic swept through the community. That Graunt was able to document these regularities was remarkable, but his awareness of the limitations of his investigations was even more impressive: He candidly revealed to his readers the inaccuracies in his data such as the inexact ages often attributed to the decedents and the frequent misdiagnoses of fatal diseases. Despite these shortcomings, he was able to make a number of important deductions about the general population and thus usher into existence the field of demographics, which deals with quantitative characteristics of entire populations. Graunt also introduced the concept of statistical samples and the actuarial tables that would lay the foundation for the insurance industry. According to Lowell J. Reed, "[Graunt] not only gave a sound analysis of this problem (the calculation of annuity prices), but he put his results in such a convenient form that this first table of mortality has remained the pattern for all subsequent tables, as to its fundamental form of expression."[3]

Graunt published his summary of his work—a tract called *Natural and Political Observations Made upon the Bills of Mortality*—in 1662. His book immediately attracted the attention of influential government leaders as well as many prominent private citizens who saw the obvious merits of documenting the demographic characteristics of the population. Indeed, King Charles II was so impressed with Graunt's work that he proposed Graunt be granted membership in the newly created Royal Society, a forum in which the nation's most brilliant scientists could gather together and exchange ideas. Even though Graunt's trade as a shopkeeper provoked some objections by some of the Society's stuffier members, the king ordered that Graunt be admitted and that admission be made available to any other tradesmen possessing such talents.

Graunt's book was distributed throughout Europe, and many continental governments began to utilize Graunt's statistical principles. Graunt's old friend, Petty, played an active role in bringing each edition of the tract to print and was even credited (erroneously) by some persons as having authored the bulk of the work. Notwithstanding Graunt's lack of support among these pro-Petty historians, however, most investigators have credited Graunt with having written the book and formulated its most fundamental principles and concepts. Petty's role seems to have been limited more to offering editorial advice because, despite the claims of his proponents, Petty did not really understand the implications of the book. Indeed, Petty seems to have been a pompous windbag who could barely carry on a cocktail party discussion about statistics and demographics.

Graunt did achieve some prominence in London society following the publication of his book. He served as a member of the city common council and was well regarded by his constituents. Unfortunately for Graunt, his subsequent decision to convert to Catholicism was not well received by some Londoners. The ensuing controversy put an end to Graunt's political ambitions and encouraged his detractors to spread malicious and unfounded rumors about him. In particular, Graunt, who had joined a water company, was accused of having ordered that the water supply be stopped just before the great fire of London started. That Graunt had actually not begun working for the water company until a month after the fire consumed the city was not so widely publicized. But Graunt weathered the controversy and spent the remaining few years of his life in London until his death in 1674. Throughout his life, however, Graunt continued to be widely admired by his friends and close acquaintances; Aubrey described him as being a faithful friend who was prudent and just. Although recognized for his discovery of statistics during his lifetime, Graunt's death went virtually unnoticed and he was buried with little fanfare at St. Dunstan's Church.

What is notable about Graunt's work is that it reflected a realization that demographic studies could be used to predict both the future growth of a population and, by inference, the future needs (e.g., food, water, housing) of that population. Graunt's statistical work would thus prove to be essential in helping governments to understand how best to manage their limited resources in view of the unlimited demands of their citizens. Many monarchs throughout Europe recognized they would have to somehow plan to deal with both the existing and future needs of their subjects if they wanted to avoid popular uprisings and the chopping block. To that end, statistics provided a very powerful tool for both identifying the needs of the population and providing a quantitative tool for helping to address those needs.

Of Populations and Samples

The Continuing Story of Mr. Statistics

O ur good friend John Graunt spent his life examining dusty church records to understand better the patterns of birth and death of the London population as a whole. It is unlikely that Graunt began his studies with a full-blown theory of statistics in his head. Instead his approach seems to have been that of a gatherer of information who did not really know what he was looking for when he began collecting copies of the *Bills of Mortality*. Whether some pivotal event in his own life or simply a strange obsession with death sparked his curiosity about the demographic landscape of London is unclear. But it seems that he may have taken a cue from his trade as a shopkeeper because his job required him to keep track of his store's inventory. Graunt, like every other merchant, had to maintain careful records of his inventory. It was only through such records that he could determine which goods were selling and needed to be reordered and which goods were of little interest to his customers. Such inflows and outflows of inventory could have provided a simple analogy to the births and deaths of the London citizenry. It may have been a natural step for Graunt to begin wondering whether certain underlying patterns could be discerned in the London population as they could be in the fluctuations of his own inventory.

But such speculations cannot be easily confirmed because many historians have not been particularly kind to Graunt. Most have come around and given him credit for inventing statistics and demographics or at least for being the individual most responsible for the creation of these fields. But they have, by and large, tended to ignore Graunt himself. The paucity of biographical information about Graunt is the most telling example of this neglect. Admittedly, his life was not one of heroic adventure or military conquest. Graunt never sailed a pirate ship or plundered a coastal village. Nor was he regarded as a supremely gifted thinker like Isaac Newton, who would retreat to the English countryside to formulate in less than two years his three laws of motion, his law of universal gravitation, calculus, and, for good measure, a comprehensive theory of optics. Graunt lived an interesting life during an age in which science was gaining the upper hand against the Church in an ongoing struggle against the religious orthodoxy that had nearly suffocated the pursuit of knowledge in Europe. Graunt also developed friendships with many important scientists, due in large part to his membership in the Royal Society. But Graunt was also somewhat plodding intellectually. His investigations were, to a certain extent, little more than laborious compilations of demographic data. He himself was not a particularly colorful individual and would have otherwise passed from history without so much as a wave of his hat were it not for the brilliant inspiration that prompted him to look for the patterns beneath the chaos of London's demographic records.

Graunt's conspicuous absence from many scientific histories may speak as much about the continuing negative perceptions many persons have about the field of statistics. Such a statement seems absurd at first glance because scientists, particularly historians of science, should know better than anyone of the important role statistics has played in the development of our technologically advanced society. But statistics is, to some degree, a discipline that requires its practitioners to roll up their sleeves and combine mathematical theory with the data yielded from detailed investigations of the real world. It lacks the sweeping grandeur of physics—which claims as its domain the physical universe from the smallest particles of matter to the most distant clusters of galaxies. Similarly, it cannot claim to govern the interactions of all the elements in nature as does chemistry. Statistics is even somewhat estranged from pure mathematics because it is not a purely conceptual endeavor like arithmetic or trigonometry. Instead, statistics is a quasimathematical, quasisocial science that must borrow from both mathematics and demographics. Despite its many real world applications, statistics continues to be saddled with the unfair (and erroneous) perception that it is tedious and of secondary importance.

What Is Statistics?

Even though Graunt's work began with mind-numbing data collections, statistics is actually much more than protracted bookkeeping exercises. What we now call statistics consists of an extensive collection of methods for designing experiments, examining data, and then analyzing the results of the experiments and drawing conclusions from those results. In other words, Graunt was forced to wear several hats. Initially, he had to think of a way to organize the mass of data revealed by the *Bills of Mortality* in a way that would reveal the actuarial features of the London population. To that end, he chose to summarize the information so that it could be more easily examined and analyzed. Graunt then had to consider the implications of the data and draw some conclusions about them. Here, he had to examine things such as the high mortality rate among children and the comparatively low numbers of individuals who died from outright starvation and come to some conclusions about these findings. The fact that Graunt's research challenged a number of popular perceptions ensured that it would be useful to city administrators. More specifically, Graunt's work forced them to consider that their own perceptions of London's problems and their prescriptions for fixing those problems might be errone-ous. So Graunt offered a new way of looking at popula-tions that involved the examination of data and the thoughtful consideration of the results of those investi-gations.

Despite its social science trappings, statistics is, in many ways, as much a science as physics or chemistry or biology. Its validity is dependent upon the care with which its experiments are carried out and the deliberate-ness with which the results are evaluated. But it—like economics, sociology, and psychology—involves study of human behavior and as such cannot claim the preci-sion of the physical sciences, which are grounded on immutable physical laws.

What is the practical significance of this difference? All natural sciences—physics, chemistry, biology—are grounded on basic physical laws that are thought by scientists to be unchanging and, as a result, totally predictable. Newton's law of gravitation, for example, predicts that an apple will fall to the ground due to the mutual gravitational attraction exerted by both the apple and the earth. We can study the effects of gravitation and make predictions about the acceleration of the apple as it plummets to the ground because the law of gravitation is a constant. This law thus makes it possible for us to predict that the apple will always fall to earth from the tree because of its gravitational acceleration. Statistics, by contrast, cannot rely on the absolute certainty of physical laws because its focus can also include animate objects such as human beings. As such it must often deal with the follies and unpredictability of human behavior.

Despite this element of imprecision, however, statistics can prove to be surprisingly accurate in the predictions it offers. But its predictive power depends on focusing on a group of events instead of a single event. Indeed, our tireless investigator Graunt soon learned that while individual events such as the exact date when a person dies cannot be predicted with any certainty, one can approximate with some certainty the average age at which an individual may perish based on an analysis of the entire population.

Suppose we want to make a fortune the easy way but we have no wealthy relatives for whom we can simply wait to die and pass to us a large inheritance. Being adverse to the idea of manual labor, we might want to try to purchase huge insurance policies on certain individuals and fervently hope they will drop dead before we have to pay additional premiums. Now we know that everyone (except the author and the copy editor) will one day die. But we do not know the precise moment at which a given individual will die unless we help the process along by dropping a boulder on the insured or stuffing a pillow over his face. Because we do not wish to spend the rest of our lives in prison, however, this is not a viable option. As a result, we must wait for nature to take its course. Unfortunately, the person whose life we are insuring could live another week, another month, another year, another decade, or another generation before we are finally able to collect the proceeds of the policy.

The actuarial tables published by insurance companies reflect their statistical analyses of the average life expectancy of men and women at any given age. From these numbers, the insurance companies then calculate the appropriate premiums for a particular individual to purchase a given amount of insurance. But the fact that the average 30-year-old man will live another 45 years does not mean that any individual 30-year-old man will live another 45 years. Some will engage in reckless activities like dating two women at the same time and die at a young age while others will manage to cling to life until they reach the century mark. Very few individuals will die at the exact age dictated by an actuarial table. So we have to distinguish between individuals who will have a wide variety of actual life spans from the gross average of all of these life spans yielded by the actuarial tables.

The point of this digression is that our plan to purchase life insurance on an individual so that we can enjoy an early retirement is fraught with the risk that the insured will live far longer than is convenient for our scheme. The actuarial tables will give us the average life expectancy of a person having the same age and gender as the proposed insured. It is not a guarantee that that person will die at that age but merely a statement of the average age at which people of that age and gender do die. So we probably will want to find some other way to make our fortune because the actuarial tables will not give us any inside information on people who are in good insurable health but destined to drop dead in a few short months.

The distinction between the ages of individuals and aggregate groups of individuals also illustrates the following principle: The greater the number of events in a group, the more likely meaningful predictions can be obtained about that group. Graunt himself would have never suggested that his statistical methods could be used to predict with any degree of certainty the age at which a London citizen would die. But by tabulating the ages at which Londoners had died over several decades, Graunt was able to discern mortality patterns that provided a great deal of accurate predictive power about the citizenry as a whole.

Because of this imperfect fit between reality and statistical predictions, statistics is considered to be inductive in nature. Its conclusions are drawn from generalizations that do not always correspond to what takes place in the real world. So long as any outcome has some possibility of occurring, such as the rather remote probability of being dealt a royal flush on three consecutive hands, it does have a certain, albeit minimal, degree of likelihood. Most poker players would dismiss such a fortuitous event as impossible, but it is still statistically possible. What about being dealt eight consecutive royal flushes? Or 15? Or 30? Even though the likelihood of such an event would become exceedingly remote, one could still calculate the statistical probability of such an event.

Populations and Samples

Statisticians often use the words "population" and "sample" when they carry out their experiments. A population is the total number of elements—for example, scores, outcomes, measurements, events—that will be studied, whereas the sample is a defined subset of the population. The usefulness of a sample is generally dependent on the extent to which it is a microcosm of the population. A good sample can help us to learn about the desired features of the entire population without having to study each and every element of that population.

Suppose you are the world's foremost expert on armadillo gaits. As a very important scientist, you must have time to do your research, teach your classes, give seminars, and present papers to your learned colleagues. Furthermore, your research takes you all around the country where you spend days, sometimes weeks, out in the wild, carrying out your research. But you obviously have neither the time nor the resources to find every single armadillo and personally study his or her gait. So you must limit the number of your research subjects and, in doing so, rely on a sample of armadillos from which you can extrapolate your findings to the entire population of armadillos. If you find that all of the armadillos in your sample study group walk with a certain lumbering canter, then you can probably assume the same to be true of the armadillo population as a whole. If, on the other hand, some armadillos lumber along while others do pirouettes, then you would be more reluctant to draw any general conclusions about the typical gait of the armadillo. Because you know that the armor-plated armadillo is not particularly well suited for whirling about on one foot, you could reasonably conclude that all armadillos lumber along without having to track down every single armadillo in the world.

Political polling provides us with a good example of the use of sampling. If we want to learn about the general public's attitude toward legalized gambling, for example, we could take either of two approaches. First, we could call up every single member of the general public and ask each individual about his or her attitude toward casinos. Although this approach would give us a very accurate result (e.g., 110,078,994 in favor, 130,223,566 opposed), such a survey would take an extraordinarily long time and cost a veritable fortune— more than the amount kept by the bank at the casino. Assume, however, that a single determined pollster spent one minute on the telephone with each of the

approximately 240 million members of the general public we have described (24 hours a day, 7 days a week, 365
days a year). Our plucky pollster would need nearly 500
years to query every single respondent. Needless to say,
the issue of legalized gambling might not then be as
pressing in the eyes of the public. A more lethargic
pollster who did not devote more than the customary
eight-hour day to this endeavor would require nearly
1,500 years to complete his investigation. So time is a
constraint that would reduce the desirability of polling
every member of the general public. The vast amount of
time needed to complete the study is another problem
because nearly everyone polled would be dead by the
time the results were tabulated—including the pollster.

As dead populations are of little value to ambitious
statisticians, the second alternative would be to try to
select a representative sample of the entire population
that could be polled quickly and inexpensively. One
way to do this would be to select a sample that reflected
the major demographic (e.g., age, sex, ethnicity) features
of the general population and then simply ask the
members of the sample groups about their attitudes
toward gambling. Of course the usefulness of this approach would depend on the degree to which the
sample mirrored the general population. If our sample
group consisted solely of chronic gamblers living in Las
Vegas, then we would have reason to suspect their
attitudes toward legalized gambling might not be truly
representative of the general population. The same conclusion would hold true if we polled only a few members of an orthodox religious sect that had recently
burned down a casino.

Statisticians must clearly be concerned with the
accuracy of their sampling techniques. But the ways in
which measurements are reported can also create misun-
derstandings about the actual results of the experiment.
Perhaps the most abused word in the statistician's

vocabulary is "average." Most of us are taught that the average value for a group of numbers, say 2, 8, 16, 16, 17, 24, and 36, can be obtained by adding all the numbers together and dividing the total by the number of terms in the group. Here we would obtain the sum of 119, which, when divided by 7, would yield an average value of 17. But statisticians, particularly those who took more than one statistics course in college, know there is more than one way to express an average. One could say 16 was the average value because it was the most common term, appearing twice. One could also say 19 was the average because it was exactly halfway between the lowest and the highest values in the group.

Unfortunately, there are certain individuals who think nothing of twisting the "average" concept to their own ends. If each of the above terms represents the test scores received by students participating in an innovative training program, the administrator of the program might point to the higher average score of 19, whereas the parents of the students who were dissatisfied with the course would likely cite the lower average score of 16. They would both be talking about the same term but they would be using it in different ways and essentially saying different things. The problem is that we as members of the public have become so accustomed to hearing about "averages" that we do not critically examine what is meant by the term. Instead we simply think of it as an expression of a common value among a group of distinct elements.

A Chinese sage once said a picture evokes a thousand words. This kernel of wisdom continues to hold true in statistics. What is particularly surprising, though, is that pictures are frequently used to misrepresent numerical data. We have all seen pictures that compare changes over time such as the growth of the average household's disposable income. If the income doubled over a period of 40 years, then an illustrator who wished to put a particularly rosy spin on the trend could draw the second dollar bill twice as long as the first. But to maintain the proportionality of the drawing, he would also have to draw the second dollar bill twice as wide as the first. The net result is that the second dollar bill would be four times as large in area as the first dollar bill and would create the erroneous visual impression that household income had actually *quadrupled* in that same 40-year time period.

Graphs can also be manipulated visually by those who wish to distract the reader from critically examining the numerical data. We have seen graphs where portions of the lines are left out so that only the most dramatic upturns or downturns are featured. One example of this would be if the unemployment rate had remained fairly steady for six years before increasing by 20 percent in the seventh year. A political candidate running against an incumbent leader might want to make a recent slump in the nation's economy a cornerstone of his campaign. He might prepare a graph that omitted the first six years of the incumbent's tenure and focused exclusively on the dramatic rise in joblessness that had occurred in the seventh year. The challenger's hope would be that the dramatic upward curve of the graph would be so disturbing that the voting public would be too upset to carefully consider the numerical data for the incumbent's entire tenure in office—even though the incumbent had six successful years in a row. In this way, the challenger would focus on the more vulnerable aspects of the incumbent's economic record.

People also use statistics to lend legitimacy to statements of fact that actually have little mathematical basis. One of the most common examples involves estimates of the number of people attending mass rallies. When several hundred thousand people show up in front of a national monument to demonstrate their support for or opposition to a particular issue—such as those favoring or opposing the death penalty—a debate will often ensue over the number of people attending. Rally supporters will often round the estimates of people in attendance upward, whereas those who do not enthusiastically embrace the cause may downplay those very same numbers. That the size of a rally can have important political consequences was demonstrated in 1995 when the controversial "Million Man March on Washington" was held. March organizers steadfastly maintained the official attendance estimates offered by the U.S. Park Service (e.g., 300,000) were too low and went so far as to threaten that government agency with legal action if it did not revise its official statement. The U.S. Park Service, for its part, ignored the threat, sticking to its estimate. Although the turnout for the rally was certainly remarkable, it fell far short of the claimed million participants and thus did not live up to its advance billing. Had the avowed goal been to bring 100,000 people to Washington, however, then the rally would have been a smashing success. So this dispute over numbers was integral to the organizers' concern over the march being perceived as a resounding success. The same point can be made about cultural events such as music festivals that draw hundreds of thousands of people from all around the country. Promoters want the public to believe these events are hugely successful, in part because it helps fuel excitement about the event and the associated ticket and "live" album sales. Here again, there is the likelihood that hard-nosed estimates of the number of people in attendance will be lacking merely because such estimates may not serve the commercial objectives of the promoters.

Of course there is a certain element of absurdity in trying to estimate the attendance at mass public rallies and concerts where tens of thousands of people are literally wandering in and out of the area in all directions at all times. Still, it is possible to come up with a reasonably good guess by taking an aerial photograph of the gathering and dividing the photograph into a grid. The estimator can count the people in a specific square and multiply that amount by the number of squares that appear to be similarly "dense" with people. Of course the obvious limitation to this approach is that not every square of the grid will have the same number of people in it. Some will be tightly packed with people while others near the edge of the crowd will be comparatively sparse. But if these variations are taken into account, then a reasonably accurate estimate is possible.

We should have less confidence about estimates when there is little opportunity to observe the entire population's size at a given moment in time. How confident can we be about the accuracy of the estimated number of fruitflies infesting a state's orchards or the number of rats living in a major city's sewers? It is simply impossible to take a snapshot of either population so they can be counted. Instead the estimators must focus on the flies found on a given piece of fruit and multiply that amount by the number of pieces of fruit on the average tree and the estimated number of affected trees.

What is the problem with such an approach? To begin with, there is no guarantee that the population of fruitflies is evenly distributed throughout the area. Indeed, they may be scattered here and there in a very uneven manner. Only if we find that hundreds and hundreds of pieces of fruit are infested with fruitflies could we then begin to come to the opposite conclusion that there was a general infestation. Notwithstanding this point, however, we would still have to be somewhat leery about calculating a number for the entire fruitfly population because we would not be able to take a snapshot of all the affected trees simultaneously.

As with our political polls, time and expense prompt such methods because the actual counting of the population would take so long that most fruitflies would be dead (and the crops destroyed) long before such an intellectually stimulating task was completed. The same issues would also apply to estimating a city's rat population. Moreover, very few people, even those who do not automatically shudder at the thought of being in the same room with a rat, are eager to descend into a sewer with a flashlight and clipboard, and begin counting. As a result, rat population estimates generally depend more on the rate at which the furry little beasts are trapped by pest control personnel and proudly held up as battle trophies in front of the television cameras than they do on any comprehensive on-site survey.

Statistics may also be used for better or worse by groups who wish to oppose a specific government policy. Suppose our nation was considering going to war against the island country of Bootsebania because their government had the audacity to serve pizza with the pepperoni under the cheese to our ambassador at a state dinner in that nation's capital. Those persons who opposed our country becoming involved in hostilities with Bootsebania might try to overstate the prowess of the vaunted Bootsebanian Navy (which consists of a fleet of kayaks stuffed with explosives) to deter our country from even thinking about engaging in war. These antiwar protesters might deliberately overcount the number of kayaks to play up in the popular press the potential casualties our country would suffer during a war with Bootsebania. Or they might simply resort to outright distortions and assert that that country's two converted crop-dusters were only a small part of Bootsebania's battle-tested air force. The point is these figures would be manipulated to serve the interests of the group, and accuracy would be sacrificed in the name of politics.

One heated political dispute in recent years has been the alleged "undercounting" by the U.S. Census Bureau of inner-city populations. In short, cash-strapped cities have repeatedly charged that the government census does not take into account all people living in these areas, in part because many of these people have no fixed address or are in the country illegally and wish to avoid detection by the government. Leaving aside the question of whether U.S. Census workers are particularly anxious to venture deeply into areas of high crime, the "undercounting" issue has become a rallying cry for many city governments because state and federal aid programs are, by and large, tied to the number of city residents. City officials have suggested that the census figures be adjusted upward to make up for the undercounting. But the obvious question then becomes "How much?" New York City could get a good deal more money from Washington if its census population was adjusted upward by 100,000 people. But it could get ten times as much state and federal funding if its population estimate was increased by one million. We can see that there is a certain financial incentive to increase the amount of adjustment as much as possible—which could conceivably outweigh the desire of city officials to obtain an accurate population count. There is no reliable way to fix a census (assuming it needs to be fixed) except by a more accurate counting system. To simply say that every urban area's population should be adjusted upward by 15 percent to take into account the "invisible residents" defeats the purpose of carrying out a census in the first place. Although the situation in every city is not the same, the use of the census to allocate federal funding will certainly encourage some city officials to try to inflate the numbers as much as possible. Similarly, we would expect rural residents to oppose any efforts to boost artificially the population counts in urban areas.

Data, Data, Everywhere

Data are to statistics what gasoline is to engines. One simply cannot engage in any kind of statistical analysis without data. We all think of data as consisting of numbers, perhaps because of our tendency to associate the word "data" with computers. But data appear in a much wider variety of forms, manifesting in many non-numerical categories. A corporation launching a new cheese-flavored onion ring boasting the minimum daily requirements of salt, fat, and cholesterol might undertake a survey of non-numerical (attribute) data such as age, gender, and ethnicity in deciding how best to target the new advertising campaign as well as how to package the new product. "Viva! Onion Rings" would appeal to the middle-class crowd, which cannot get enough nachos and salsa, whereas "Farmer Brown's Old Fashioned Onion Rings" might be of greater interest to the vegetarian or "down home" group. Alternatively, our marketing research might conclude that an approach that dared America to have the guts to eat this product would be preferable, thus prompting the name "Fat-Filled, Lard-Laced, Artery-Clogging Onion Rings."

Our beloved political pollsters are also fond of compiling data on the personal characteristics of their respondents because it helps them analyze how well their samples represent the whole population. Of course, this tendency to focus on relatively unimportant sample characteristics could be unproductive. It might not even be too far-fetched to imagine a television broadcaster reporting election night returns and noting with the greatest degree of seriousness that "women with red hair have voted overwhelmingly for the incumbent" whereas "the challenger continues to draw much of her support from the leprosy community." Obviously, this type of reporting would not really give us any great insight into the reasons which prompted people to vote the way they did.

When we talk about numerical data, we may speak of either discrete or continuous data. Discrete data consist of a finite or countable number of values, whereas continuous data may have an infinite number of values. If we count the number of milking cows at several local dairies, we will come up with discrete data because we will be able to say how many cows (e.g., 5, 10, 20) are at each of the dairies. If we actually tried to measure with infinite precision the amount of milk each cow produced in a given day, we might obtain quantities that could extend anywhere between 0 and a certain whole number but also include any or all the numbers (decimals) in between those two points. In other words, Agnes, a truly motivated milking cow, might give 7.345 quarts of milk on a typical day. But there will be some days when Agnes is depressed from having contemplated the dreary pointlessness of her own life, what with its unchanging routine of grazing and milking. On one of these "off days" when Agnes's enthusiasm for mechanical milking machines is at a low point, she might give only 2.076 quarts of milk in one day. If we chose to measure this amount with greater precision, we might find that Agnes had given 2.076582 quarts of milk that day. The point is that we could have any of an infinite variety of measurements for a particular cow's milk production, depending on how accurately (to how many decimal places) we wished to measure the volume of that cow's milk production.

There are several different types or levels of measurement that are used by statisticians to order data. The simplest level of measurement, though not terribly helpful for carrying out statistical studies, is the nominal level of measurement. This level provides no basis for statistical comparisons because it simply compiles lists of names. A newspaper article tallying how many Republicans and Democrats voted in favor of or against a law to ban smoking in government buildings would be a nominal level of measurement. The fact that 39 Democrats and 23 Republicans voted in favor of the measure would not be of great interest by itself to statisticians because such information does not have any numerical or comparative attributes. We cannot draw any meaningful conclusions from this tally because the data cannot be ordered. Even if we assign an arbitrary value to each legislator's vote—ostensibly to gauge the political activism of our elected representatives—we cannot make any calculations that have a meaningful quantifiable basis because the way in which we choose to assign these values is not rooted in any general standard of measure.

An ordinal level of measurement may be used to arrange data in some order, but such arrangements typically lack enough information to evaluate the differences between the values. If a group of ten schoolchildren rate leading breakfast cereals for their ability to stay crunchy in milk and use nonquantitative rankings such as "poor," "fair," "good," or "excellent," then they would be using an ordinal level of measurement. The fact that one cereal such as Wheat Chips was rated as "poor" because it turned to mush as soon as milk was added to the bowl whereas another cereal such as Oat Compost was rated as "excellent" because it stayed crunchy in milk for weeks would not, by itself, be very helpful, because there would be no obvious quantitative differences between the ratings "poor" and "excellent." In short, these ratings alone would not provide us with enough information to determine the quantitative distinctions between the rankings or make a determination as to which cereal was preferred by the people participating in the taste test.

If we were able to determine in some meaningful way the differences between the data, then we would be using an interval level of measurement. A thermometer, for example, provides us with quantitative differences between temperatures. We know immediately that a temperature of 100 degrees is 40 degrees higher than a temperature of 60 degrees. As a result, we can arrange the values of our thermometer in an order that provides us with meaningful differences. But the interval level of measurement is not without its drawbacks; it does not provide meaningful information about ratios because it does not have an ultimate minimal value. A measurement of 90 degrees Celsius, for example, is not three times as warm as a measurement of 30 degrees Celsius. Moreover, a measurement of the temperature at which water freezes (0 degrees Celsius) does not represent the lowest possible temperature in nature but is instead a numerical value that is arbitrarily assigned on the Celsius scale. There are much lower temperatures in nature that can be experienced by any visitor to Chicago during the winter months or a husband who forgets his wife's birthday. Instead we must turn to the Kelvin scale of temperatures, which offers an absolute zero temperature, to obtain information about ratios in terms of different measurements of heat. Unlike the Celsius scale, the Kelvin scale provides a measure of the coldest possible temperature in the universe and thus permits us to draw meaningful conclusions about both differences and ratios. So the Kelvin scale provides a ratio level of measurement whereas the Celsius scale, due to its lack of an absolute zero value, can provide only an interval level of measurement of differences.

Because statisticians live for precision, they will prefer ratio levels of measurement to interval levels of measurement. But they will make do with interval levels of measurement if the only other alternatives are the correspondingly less quantifiable ordinal and nominal levels of measurement. This pecking order imposes certain constraints on statisticians because they cannot use the data obtained from using a higher level of measurement (e.g., ratio or interval) to draw meaningful conclusions on a lower level of measurement (e.g., nominal). In other words, it makes no sense to apply the concept of "intervals" or "differences" to data consisting solely of compilations of names. For example, we cannot make a meaningful statement about the ratio of Ralph to Janet because we are dealing with nonquantifiable elements (proper names).

The Art of Sampling

Much of the value of any statistical analysis depends on the methods used by the statistician to design and carry out the experiment. These methods will in turn depend on how the person carrying out the experiment formulates the objective of the experiment. As with anything else, a clearly stated purpose is preferable to one lacking clear direction. The Manhattan Project, which originated when Albert Einstein sent a letter to President Franklin Roosevelt during World War II, was guided by a single overriding objective: to construct an atomic bomb before Germany did. Although a typical statistical study is much more mundane than a top-secret government research project, it is preferable that it also be guided by a single purpose. Otherwise, the experimenter may tend to become mired in the minutia of his data or become careless or sloppy in his experimental methods. This is not to say that experimenters do not have their own personal reasons for carrying out a research project in a given way or that these very same experimenters do not have biases that may cloud the interpretations that can be drawn from their research. It is clear, however, that the carrying out of a scientific experiment is greatly facilitated when the experimenter is seeking answers that are not clothed in subjective garments.

Once the purpose of the experiment has been decided upon, the process by which the data will be collected must be designed. By design, we are referring to the plan by which the data itself will be collected for the experiment. This design is extremely important because the experimenters want to make sure the data they are going to so much trouble to collect will actually answer their questions or at least be the type of data that can help to resolve these questions. At this point, they will have to determine whether a given sample of the population being investigated contains enough information so it can be viewed as being truly representative of the population. Otherwise, the sample will have little value for purposes of drawing conclusions about the entire population. A poorly designed study will be discounted, if not ridiculed, for its flaws.

Sampling enthusiasts will be pleased to discover there are a number of different methods by which valid samples may be obtained. Random sampling is perhaps the most appealing approach because it presupposes each member of a population has an equal chance of being selected for inclusion in the sample. The devices used by state lotteries, for example, by which five, six, or even seven numbered balls are selected to give the winning lottery numbers illustrate this concept of randomness. The success of these lotteries clearly depends on the public perception that every number has an equal chance of being selected. Of course, the selection of lottery numbers is not intended to serve as part of a statistical experiment but to award several million dollars to a lucky winner. The important point is that every numbered ball, over a reasonable period of time, should (in principle) be selected as many times as every other ball. The lottery machine has an advantage over most populations featured in statistical experiments in that the ball population is completely defined, and the samples are merely different combinations of the same population of balls. The balls all weigh the same and each has one unique number on it. All of the balls are made of the same substance, and no ball has a sticky surface that might otherwise give it an extra advantage. We do not have a nonparticipation issue as may be the case when surveys are conducted and at least some of the persons asked to respond to questions refuse to do so. This is because we are concerned with uncovering the attitudes of the population as a whole; the refusal by many of the respondents to participate in the survey makes it impossible for us to be completely confident about the survey's accuracy.

We can better understand this problem of potential bias by considering the plucky political pollster who stands on a streetcorner asking passers-by about their opinions of a recently enacted controversial ban on the production, sale, or consumption of onions. Some people, particularly those who might have been trapped next to onion lovers on public transportation, might be very enthusiastic about the ban. Others, such as the consumers of onions, might be vehemently opposed to this ban, citing their constitutional rights to consume malodorous vegetation throughout the United States. But there could be other people who would not want to voice their opinion to the pollster, such as monks keeping a vow of silence or those people who simply do not want to take the time to stop and talk with the pollster. If the number of nonrespondents is significant, then the pollster will have to be concerned about the accuracy of the data because nonresponses erode the accuracy of any sampling of public opinion. Furthermore, the pollster will have to be concerned with the reasons why so many persons are refusing to respond to the questions. If the reason is simply a lack of interest in standing on a busy sidewalk with a pollster who wears polyester suits and wide ties, then the lack of response does not necessarily relate to the onion ban issue. But if the lack of response is because most of the nonrespondents belong to a secret onion cult and do not want to open their mouths in the presence of strangers with camera crews, then it is probable that the pollster's results regarding public support for the ban will be inaccurate. We also need to remember that there are people who refuse to speak to pollsters because they worry their neighbors will somehow find out they are closet onion lovers.

Researchers also use stratified sampling techniques whereby they divide respondents into distinct classes based on certain characteristics such as age or gender. Suppose a legislator with an elderly constituency proposed a bill that would impose a "youth" tax on anyone under the age of 65 so that annual $10,000 payments could be made to anyone over the age of 65. A pollster might divide the population into two groups (under 65 years old and 65 and older) and then ask random samples within each group about their attitudes toward the bill. This stratification would give us more information than would otherwise be obtained from a simple random sampling of the population of the whole because it would show the support for the bill among two distinct groups—those who would benefit from it and those who would pay for it.

What approach could be used if you wanted to get a good sample from the membership body of a massive organization such as the American Association of Grumpy Persons? You could purchase a copy of the membership list, pick a name anywhere in the directory and then skip ahead and select at regular intervals (for example, every 30 names) a new name for your sample until you had covered the entire list. With a group having tens of thousands of members, a systematic sampling process that culled every 30th name after a random starting point would provide us with a simple, effective method for randomly sampling a large population. Of course such a sampling process could not guarantee the respondents would be sufficiently grumpy or even arrogant, but such things are often beyond the control of even the most adept statistician.

Elections are a fertile source of employment opportunities for pollsters. Political pollsters often conduct polls to determine the general public's attitude toward people who might be running for public office. Yet the need to gauge public opinion can create difficulties, particularly when dealing with a national campaign. After all, a nationwide poll is a massive undertaking, and it would be an exhausting task to travel to every city and town to sample the local population's attitudes toward the leading candidates. This task is further complicated by the fact that the attitudes of at least some of the respondents will change over time. One way out of this predicament is to use cluster sampling and to divide the electorate into a collection of distinct areas such as counties or towns and then sample the prevailing opinions in a few select areas. Hence, the travel-weary pollster would see a way to avoid a polling expedition that would take him to every valley and port of this country by simply picking a few cities, towns, and rural areas that he believed would accurately represent the country as a whole. He could then intensively canvas the public in those few areas and perhaps develop more thorough and accurate polls. In any event, this cluster sampling technique would spare him from having to climb aboard a plane every few hours for the next several years.

The pollster who prefers the softer life can use convenience sampling, which merely means that he does not exert himself too much to obtain the sample. He might simply wander into a grocery store and bar the door until each person there agreed to answer his 30-page questionnaire detailing their opinions of the Federal Reserve's management of the nation's money supply. The problem with convenience sampling is that it can be too easy. Suppose our lollygagging statistician asked those very same people about their views of the wisdom of a sales tax on food. Given that these shoppers all want to stretch their food dollar, it is unlikely that the patrons would break into rousing cheers when offered the tax proposal. No, the statistician would have to find a more neutral ground to ask questions about food sales taxes. But he would have to keep in mind the potential for biased responses. Finding out how a proposed food tax would affect the respondent's own welfare would certainly help the statistician to be alert to possible biases. Most people as consumers of food would probably have little enthusiasm for a food sales tax because it will raise the cost of food. However, a government official whose job was dependent on the state being able to obtain more tax revenues would probably be more enthusiastic about the proposal.

Despite the wide variety of sampling procedures available to pollsters, no method is foolproof. Indeed, the accuracy of these methods is heavily dependent on the ways in which the pollster's questions are formulated. Such nonsampling errors, which may involve the use of subjective, inflammatory, or vague questions to evoke emotional—as opposed to rational—responses can scuttle the most precise sampling methodology. Needless to say, some pseudopollsters might go so far as to couch rumors in terms of questions if only to sabotage the political campaign of an opponent. But if one is interested in merely gauging the public's attitudes toward a certain candidate or issue, it makes little sense to spend great quantities of resources designating wonderfully representative samples when those sampled will be asked questions that are biased or unclear.

Obvious and Subtle Statistics

Factual science may collect statistics and make charts. But its predictions are, as has been well said, but past history reversed.
—JOHN DEWEY

Introduction

A s we can do what we want in this book in the interest of conveying the basic concepts of statistics and illustrating some of the more common applications, we shall now move forward and try to learn something about the animal we call statistics. There are two distinct approaches used by statisticians to tackle questions they may have about a given population. Descriptive statistics is the most basic form of quantitative analysis because it essentially summarizes or describes the characteristics of a given set of values in a population. If we have a list of the heights of some of the football players on a team, for example, we would use descriptive statistics to summarize or describe this information (the heights of the players in the sample). As these data are already laid out for us, we can think of descriptive statistics as being "obvious" statistics. In short, it requires no additional effort on behalf of the analyst to draw any conclusions about the information. As such, the importance of the numbers is obvious.

What is "subtle statistics?" We refer to inferential statistics as "subtle" because it goes a step beyond listing numerical data from a sample such as the heights of our football players and, in addition, requires us to draw conclusions about the entire population (team) of football players itself based upon the information contained in the sample. If the average height of the ten football players listed in the sample is six feet even, then our conclusion that the average height of all the players on the football team is six feet even would constitute an inference about the population based upon the information contained in the sample.

Remember, we are using words such as "values" and "population" as they would be defined by statisticians. As a result, we would not, for example, define the word "value" as pertaining to social mores or the bargains to be found if one is willing to endure the feeding frenzy going on during the spring sales at the local department stores. No, these words must have some numerical aspect because statistics is, after all, a field in which numbers are everything.

When we grapple with descriptive statistics, however, we take on the collection or values of numbers directly. If we were given 200 scores (values) that represented different numerical grades (ranging from 0 to 100) received by students in the shop class at an elite private school, then we could describe these values simply by listing all the grades in a single table. For good measure, we could certainly press ahead and calculate the average numerical grade by adding the total grades and dividing by the number of students receiving each grade. If the average grade was 88, then we could feel quite confident about the future of the United States' machine tool industries. Indeed, we could walk with a swagger in our step and a smug smile on our face whenever we journeyed abroad, knowing that our nation had the upper crust of the world's future shop class graduates. However, if the average grade was 44, then we might want to begin learning to speak German or Japanese.

This calculation of the average value would not really tell us very much about the student population as a whole unless there were only 200 total students who took shop class. Suppose we still had only the 200 grades but there were actually 800 students who took shop class at that school. We would then have to wonder whether our average grade was indeed truly representative of the student shop class population as a whole. The 200 grades would thus constitute a sample equal to 25 percent of the total population. Now this would be a very significant sample size because statisticians are often forced to rely on comparatively puny sample sizes that may be no more than a fraction of a percent of the total population. But once we made the justifiable assumption that the average value of the 200 grades would be a good approximation of the entire 800 grade population, then we would be moving into the realm of inferential statistics. We would thus be using a sample size (of considerable size relative to the entire population) to make inferences about some characteristic (grades) of the population as a whole. In short, the sample size would provide us with a basis for extrapolating our findings or conclusions about the cutting-edge shop class education being received by all 800 students.

Although inferential statistics constitutes much of what we regard as the field of statistics, we must become better acquainted with descriptive statistics. No matter how much students beg and plead their statistics teachers to plunge into inferential statistics, it makes more sense for teachers to begin with descriptive statistics; it is much simpler conceptually to understand as it is merely a given set of numerical values. Moreover, its shortcomings as a conceptual tool will become apparent very quickly and lead us quite naturally to inferential statistics.

We do not want to create the impression that descriptive statistics is a sort of halfway house for wayward statisticians. No, descriptive statistics is both the beginning and end of statistics because it is conceptually simple yet demands all knowledge about the population being studied. For this reason it is the end to which all statistical studies are geared. It is also, in many cases, an impossible goal that cannot be reached because the individuals carrying out the research simply do not have the time, energy, or resources to describe statistically the data population they are examining.

Frequency Tables

In those cases where we do have enough information available to describe the values for the entire population, we can use what is called a frequency table to provide us with a snapshot view of the entire population. To return to our example of the shop class students, we could construct a frequency table that would consist of ten horizontal columns numbered consecutively in the following manner: 0-10, 11-20, 21-30, and so on. To the immediate right of each of these ten columns we would record the number of grade averages that fell within each of those ten columns. So if there were 18 students at the school whose average grades were between 0-10, we would put the number "18" to the right of the 0-10 column and prepare them for careers in politics. And if there were 35 students who had studied late into the night to master their hand sanding techniques and posted lofty averages between 11-20, we would place the number "35" to the right of the 11-20 column. By repeating this step for each of the ten horizontal columns, we would construct a frequency table that would tell us the distribution of grades for all of the 200 shop students:

Score	Number of Students
0-10	18
11-20	35
21-30	16
31-40	25
41-50	58
51-60	33
61-70	8
71-80	5
81-90	1
91-100	1

But there is a price that the frequency table will exact on our shop class student data. Although we would have originally begun with all 200 numerical grades of the students, we would now find ourselves with a ten-column chart that would tell us only that a certain number of grades fell within each of those ten columns. In other words, we would no longer have the specific scores for all 200 students in this summary. Even if only one student scored in the 91-100 range, we would not be able to determine his score by looking at the frequency table because the table would merely tell us one student had scored in that ten point range. (We could probably track that student down in person by counting the number of fingers on the hands of the students in the power saw class.)

Because the frequency table is a compilation, it necessarily simplifies the mass of data (the 200 scores) that we were previously trying to organize. What we lose in accuracy, we make up for in ease of use and comprehension. A good frequency table will have a fairly small number of classes (like the ten columns tabulating the shop class grades) that have distinct boundaries (do not overlap). Anarchists who do not believe there should be any boundaries do not make good statisticians because one cannot construct non-mutually exclusive classes and still obtain useful frequency tables.

A statistician with some spare time on his hands can convert a frequency table into a cumulative frequency table merely by successively adding each of the quantities associated with the various ranges as he progresses down the page. To return to our shop class example, we would still place the number 18 after the first column, 0-10, because we saw that 18 would-be valedictorians received a grade between 0 and 10. But the second column, 11-20, would combine not only the 35 near-geniuses who received a grade between 11 and 20 but also the 18 students in the previous column for a grand total of 53. We would then continue this process as we moved up the shop class food chain to those few students who did not have staple-gun scars on their hands or feet.

A different way to look at the proportionate representation of each of these classes is to create a relative frequency table. In this example, we begin with the 0-10 range and divide the number 18 by the total number of grades in the table (200) to get a relative frequency percentage of .09 or 9 percent. Similarly, the 35 students who scored between 11 and 20 would represent 35/200 or .175 or 17.5 percent. The advantage of a relative frequency percentage is that it provides a very accessible breakdown of the distribution of values in a population (in this case, the grades of the students in shop class). In other words, the relative frequency distribution table would provide a column-by-column breakdown of the relative percentages of each grouping of values. It would be a little like taking the 100 pennies that make up a dollar and putting 9 pennies by the first column, 17.5 pennies by the second column (we could chop a penny in half), and so on.

Picture Graphs

Not all people are enamored with numbers. Even statisticians will admit that pictorial graphs can be extremely effective in communicating information about certain characteristics of a population. One of the most common forms of picture graphs is the pie chart, so named because it typically consists of a circle divided into different "slices" to graphically convey the desired information. The ease with which a pie chart accurately conveys information is critically dependent on the accuracy of the proportions shown on the chart. In other words, there must be an exact correlation between the proportionate distributions of the data and the rendering of the chart itself. A pie chart drawn to show that 38 percent of all automobile accidents are caused by "stupid drivers" would necessarily have a slice bounded by 38 percent of the circumference of the pie circle (which on a 360 degree circle would be equal to 136.8 degrees of arc or 38 percent of the circle's arc), as shown in Figure 1. A second slice showing that 19 percent of all automobile accidents are caused by "lap dogs that suddenly turn on their masters while they are driving the car" would be bounded by 19 percent of the circumference of the pie circle (and equal to 68.4 degrees of arc on the circle), and so on for each additional slice of the pie circle. But the advantage of pie charts (the ability to show proportions) also brings certain disadvantages, such as the loss of detailed information about all the values used to calculate the relative sizes of the pie slices. In other words, we have only the total number of accidents caused by

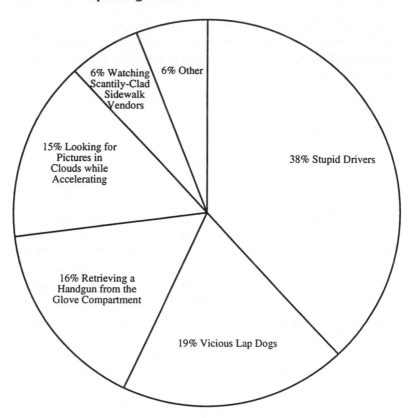

Figure 1. Pie chart describing the major causes of auto accidents.

stupid drivers, vicious lapdogs, gun-toting drivers, cloud watchers, and daydreamers. We have no information about the individual incidents that figured into each of these totals. In fact, we do not even know the total number of accidents that resulted from each of these situations. As a result, we have no idea how many hundreds, thousands, or even tens of thousands of accidents are being described by this pie chart.

Frequency tables have a graphical counterpart known as histograms or bar charts. In other words, we can convert frequency tables, including cumulative and relative frequency tables, into graphical form merely by placing the frequencies along one axis and the values to be found at each of those frequencies along the other axis and then drawing bars that rise upward from the horizontal axis to the appropriate level of the vertical axis. These bar charts are conceptually similar to pie charts in that they represent distributions of values. They also have the same limitation because the amount of detailed data must necessarily be minimized.

How might the grades of our shop school students be depicted in a histogram? We could have the grades (0-10, 11-20, 21-30) on the horizontal axis and the number of students (10, 20, 30) along the vertical axis, as shown in Figure 2. We could then construct a bar chart for our shop student frequency table by drawing the first bar so that it rises from the 0-10 column upward until reaching a level on the vertical axis that would seem to approximate 18 students. This step would be repeated for each successive column until the grades of

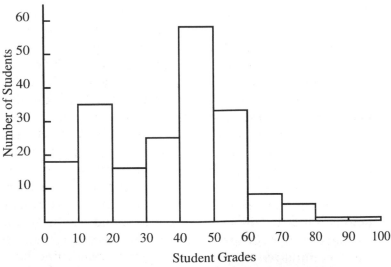

Figure 2. Histogram describing distribution of student grades.

all 200 students were represented. A cumulative frequency bar chart could be constructed merely by piling the additional incremental increases on top of each successive bar with the final bar rising all the way up to a value of 200 (for the 200 students who are in shop) in the final column. Similarly, a relative frequency diagram could be constructed merely by substituting percentages (9 percent, 17.5 percent, and so on) on the vertical axis and then drawing the appropriate bars.

Must the bar chart be the sole refuge for truly visually oriented statisticians? No, indeed. We can save a great deal of time and ink by merely placing a dot at the appropriate level above each horizontal column or class (so that the first dot would be placed at the number "18" on the vertical axis above the 0-10 grade value of the horizontal axis and the second dot would be placed at the number "35" on the vertical axis above the 11-20 grade value of the horizontal axis, and so on). We will eventually obtain a series of dots that can be connected to create a line graph. If we place each dot so that it corresponds to the number of values in a given class (for example, 18 for the 0-10 column, 35 for the 11-20 group, and so on), we will create a frequency polygon. If we graph the cumulative frequencies, however, then we would still place our first dot at 18 for the 0-10 column but we would then place the next dot at 53 (18 + 35) for the 11-20 column and so on. This cumulative frequency polygon will usually create a sort of reverse "S" pattern as the dots over each of the classes are connected. A cumulative frequency polygon of the grades received by our shop students would be useful for showing how many students scored below or above a certain score. So the hiring coordinator for a tool and die plant who is interested in securing the best and the brightest workers for her company's assembly line might want to use a cumulative frequency polygon to determine how many students finished in the top ten percent of their class.

Means, Medians, and Modes

Frequency tables and pictorial graphs all illustrate how specific values of a given population are distributed. They can also provide us with information as to the average value of the population. Before we plunge ahead into the morass, it is important to point out that the word "average" is somewhat misleading because there are several different, but perfectly acceptable, ways to calculate the average value of a population. These different calculations lead us to different averages, which are known as the mean, median, and mode.

The mean value of a population is the one most familiar to people who do not read such weighty tomes as the *Statistical Abstract of the United States*. It is very simple to calculate because you merely add up all the numerical values of the population and then divide by the number of values. So if I want to calculate the average number of patients who visit the local plastic surgeon each day, I would need to determine the total number of visits that occurred in a given period of time (such as a month or a year) and then divide that total by the number of days being considered. Whether these patients wanted to have their buttocks lifted to their shoulders or their noses whittled down to the size of a bottlecap would make no difference in our calculation because we are concerned only with the total number of visits. As a result, we might count the entries in the appointment books for an entire month and divide that total by the number of days in the month. This quotient would give us the arithmetic mean. The arithmetic mean (which we shall refer to as the mean) is nothing more than an indication of the central values around which the population (in this case, the number of visits per day) as a whole clusters. So even though there may be an average value of 22 patients per day visiting the plastic surgeon, there may not have been a single day in which exactly 22 patients walked into the office. In other words, the numbers may have ranged as high as 44 on the day when a minor earthquake caused many face-lifts to sag, sending patients scurrying for reconstructive assistance to as low as 0 on the day when the surgeon was called to the golf course for an emergency skins game. But the mean does provide us with some idea of how many patients on average can be expected to saunter up to the receptionist and inquire as to when the doctor will be able to see them. How could the doctor use this information? First, he would get some idea as to number of people who venture into his office each day and, by calculating the total billings, the amount of revenue generated each day by these patients. Armed with this information, he could make all sorts of strategic decisions such as which days are less profitable (and thus lend themselves to additional rounds of golf) and which days are so busy that he must have an additional nurse on site.

The mean is only the *most common* type of average used by statisticians. Another measurement, the median, provides additional clues as to the distribution of values in a population. The median is obtained by arranging all the values of a population in order of ascending magnitude and then picking the middle value. So if our plastic surgeon was attempting to find the median value of the number of patients visiting his office in a single week, he would arrange the number of patient visits for each of his five office days (18, 18, 19, 21, 23) and thus determine the median is 19 because there are an equal number of values both below and above it. What if he decided on a lark to keep his office open a sixth day that week so that the number of patient visits for each of the six days that week was as follows: 18, 18, 19, 21, 23, and 25? Because of our having an even number of values, we would add the middle two numbers in this ascending series (19 and 21) and divide that number by two to get the median value of 20. This median value in this case is merely the mean of the two middle numbers (19 and 21) in our ascending series.

Our surgeon might also want to find the mode value for the number of visitors to his office. The mode is simply the value that occurs most frequently in a given population of values. If we consider again the number of patients who came to the office during the six-day work week, we would see that 18 patients visited the office on two separate days. Because different numbers of patients visited the office on each of the other four days, 18 would be the mode value. Knowing the mode would enable the surgeon to plan his office hours and determine his staffing requirements; the mode value, whatever its quantity, provides a sort of overview of the flow of patients into the office. Of course our surgeon might not want to rely on a single week of patient visits when determining how many nurses and administrative personnel to hire. Instead, he might calculate the mode value for daily patient visits over a three- or six-month period so that he would feel more confident about his conclusions.

Weighted Means

Our earlier ventures into the wonderful world of averages implicitly assumed that all of the numbers being considered (whether they are student grades or patients) are equally important in the calculation of the mean, median, or mode. But not all numbers are created equally. Nowhere is this more apparent than in the dusty lecture halls of America's great colleges and universities, where students soon learn that not all grades are created equal. In short, most college courses require the students to complete a mix of papers and examinations before they receive their final grades. Not surprisingly, most professors weigh the final examination more heavily than the midterm quizzes or reports. Such an approach has a certain logical appeal because the final examination is typically a comprehensive review of the entire course whereas quizzes are usually much more limited in scope, covering only a part of the course materials. This weighing may be of little comfort to students who embrace final examinations with as much enthusiasm as they would members of a leper's colony. Different tests counting for different amounts in the tabulation of the final grade is an inescapable part of the modern university.

How would you calculate a weighted mean? It is really quite simple even though it has never been a favorite party trick. We can begin with a student (Bob) who sat through an introductory economics class because he wanted to be conversant in topics regarding macroeconomic policy and, perhaps, meet the lovely redheaded woman in the second row of the lecture hall. Because Bob endured four quizzes, each worth 15 percent of his grade, and a final exam worth the other 40 percent, his final grade will depend on the relative weights of his various test scores. If Bob scores 85, 77, 86, and 96 on his four quizzes and 88 on his final exam, his instructor would calculate his final grade by multiplying each grade by the relative importance (percentage) of the grade and adding all of these numbers together. Then he would divide this number by the total of the weights. Our budding John Maynard Keynes would have the following weighted mean: $(85 \times 15) + (77 \times 15) + (86 \times 15) + (96 \times 15) + (88 \times 40) = 8680$ which, when divided by 100 (the total of the weights—$15 + 15 + 15 + 15 + 40$), would yield a weighted mean of 86.8. So Bob would be able to hold his head up high, knowing he had received a B or B+ in this challenging course. Whether this incredible demonstration of intellect would be enough to win the heart of the object of his affections would remain to be seen. Even if Bob were spurned or spat upon, he could still take pride in having performed strongly on the most critical and, hence, pressure-packed, part of the course—the final exam.

Statistics for Deviants

In our examination of means, medians, and modes, we were concerned with different ways of expressing a central value or standard. These types of averages, useful as they may be, cannot tell us a great deal about the distribution of values (e.g., grades, prices, patients) in a population. They can provide us with some insight about the "clustering" of a population in a given interval of time but they cannot tell us much about the greatest and smallest values in that population or how likely it is that these values will occur over time. In short, these types of averages give us no real clue as to the spread of the values of the population as a whole because they can only provide a single value.

Fortunately, statisticians have thought of everything in developing dispersion statistics to measure the amount by which the values in the population are scattered around the central value (the mean, median, or mode). In particular, there are three measures of this scattering known as the range, standard deviation, and variance. In a sense, these values are the three counterparts to the mean, median, or mode. So while the mean, median, and mode provide us with some idea as to how the values cluster, the range, standard deviation, and variance show us how these values are distributed. We shall see how each of these three measures of dispersion manifest in statistical analyses. Although this revelation is not quite as profound as discovering the meaning of life or the theory of relativity, it does illustrate that statistics can boast several different ways by which both the clustering and distribution of values can be measured.

To better understand the usefulness of these measures, we have to understand the measures themselves. The first measure of dispersion is called the range, which has nothing to do with the Great Plains. It is, at its basic level, a measure of the extent to which a collection of values (usually numbers) are spread out. Suppose we wanted to find the range of values of the mortgages held by Millstone Mortgage Company. Our review of the company's records would reveal that some of the mortgages are extremely small in value whereas other mortgages are very large. If our exhaustive investigations (presumably not done at night while holding flashlights and wearing ski masks) showed that the smallest of Millstone's mortgages is $34,000 whereas the biggest is $2,345,000, then we could easily calculate the range of the mortgages by subtracting the smallest mortgage from the largest. Our crack mathematical minds would quickly conclude the range of the value of loans held by Millstone was $2,311,000. Unfortunately, this information does not cause us to leap in ecstasy or shout "Eureka!" or even roll over in bed because the range does not really tell us much about the population (in this case, the mortgages held by Millstone) other than the difference between the largest and smallest values.

The range's limitations as a measure of dispersion are further shown if we compare Millstone's collection of mortgages with those belonging to its bitter competitor, Really Friendly Bank. Suppose the range of the value of loans made by both banks is about two million dollars. Millstone makes a variety of loans of differing amounts to homeowners while Friendly concentrates primarily on loans ranging from $100,000 to $500,000. The fact that both banks have loan portfolios with the same ranges would obscure the fact that they pursue wildly different strategies regarding their homeowner loans.

While the range can provide us with only a superficial view of the distribution of values (e.g., mortgage loans) in a population, we need to take a deep breath and try to derive the standard deviation. One reason the range is so popular is that it is very easy to obtain. We merely have to subtract the lowest value from the highest value. As a result, it is very appealing to couch potato statisticians because it does not take very much time away from their television viewing schedules. But the standard deviation is where we begin to separate those truly intent on learning statistics from those otherwise considered to be "numerically challenged." The standard deviation is perhaps the most important measure in statistics, but it can be somewhat cumbersome to obtain. Of course, the task of deriving standard deviations has been made easier with pocket calculators, but the formula for the standard deviation can appear daunting at first glance. As with any other formula, however, the trick is not to allow your eyes to glaze over when first presented with the formula but to focus on its components and to break it down, part by part. In the same way that a raging bull can be reduced to a harmless collection of steaks, the standard deviation can be understood by approaching each component of the formula one step at a time.

The actual formula for determining the sample standard deviation is the following equation:

$$\sqrt{\frac{\Sigma\,(x - \bar{x})^2}{n - 1}}$$

Most people are surprised to find that deriving the sample standard deviation for a population can be less painful than severing one's legs with a chain saw or even stubbing a toe on the leg of the living room couch. How do we decode this formidable looking formula? Simply put, we begin at the beginning. First, we go to our scores—whether they be grades, prices, or whatever—and find the mean. That is, we add up all the scores and divide by the number of scores. The value of each individual score is represented in the formula by the symbol x and the number of scores is represented by the symbol n. We subtract the mean, represented by the symbol $\bar{x}$, from each of the individual scores in the population: $(x - \bar{x})$. If we were trying to determine the standard deviation for a population of student grades, we would subtract the mean value from each student's grades. After taking a break to recover our breath, we would square each of the differences obtained from the previous step: $(x - \bar{x})^2$. We would plunge ahead and add all of the squares obtained from the previous step to get $\Sigma(x - \bar{x})^2$. Having crossed the barren wastelands of basic mathematics, we would then divide this total by the number $(n - 1)$, which, in a population with ten values (grades), would be equal to $10 - 1$ or 9. Finally, we would find the square root of the result just as the sun disappeared beyond the horizon.

Whew! Fortunately, most calculators contain function keys that permit even the least enthusiastic of statisticians to obtain sample standard deviations for a given population in a matter of seconds. The wonderful thing about this formula is that it also contains everything you need to know to find the variance of the same population. How can this be? you might wonder. Quite simply, the variance is found by following the same procedure used to determine the standard deviation. However, we obtain the variance by omitting the final step of the process. This means the variance can be obtained with the following formula:

$$\frac{\Sigma(x - \bar{x})^2}{n - 1}$$

In other words, we do not find the square root of the result that would otherwise be gained by following the procedure described in the previous paragraph. The reader obtains two very impressive statistical formulas for the price of one and can now one-up most of the people with whom he or she will converse at the next cocktail party. For the present time, however, we will concentrate on the standard deviation, leaving the variance for later discussion.

What is the point of finding the standard deviation of a population? Well, it enables us to determine the uniformity or consistency of values—including grades, scores, prices—in a population. This may not seem like an important thing, but every company selling products in the marketplace is vitally interested in product consistency. After all, a doll manufacturer, for example, wants to make sure every one of its dolls has two arms, two legs, and one head. Unless it is planning to market a mutant line of dolls with five, six, or even seven limbs, the company will want to control its production process and, hence, reduce the variations in its products as much as possible.

Can one determine the standard deviation of a population in the privacy of one's own home? Absolutely! Indeed, one can spend hours and hours selecting populations and calculating standard deviations for those populations. Although such a habit might suggest the existence of deeply rooted personal problems, it does enable even the most timid student of mathematics to become very adept at examining the distribution of values in populations.

Remember, the golden rule for understanding the standard deviation of a population is: *The greater the standard deviation, the more the values within the population vary from each other.* But talk is cheap. We must calculate the standard deviation for a population to better illustrate the simplicity of this seemingly cumbersome process. Suppose we wish to calculate the standard deviation for the number of cars owned by four extremely wealthy families living in the extraordinarily exclusive develop-ment known as Snobby Ridge. By sneaking over the stone wall and avoiding the half-blind guards and the rabid watchdogs, we could count the number of cars kept at each of the four mansions. Our arithmetic would reveal that the four families owned 4, 5, 6, and 9 automobiles, respectively. We would then begin our search for the stan-dard deviation by finding the mean (4 + 5 + 6 + 9 = 24) and dividing 24 by 4 to get the mean value of 6. By sub-tracting the mean value of 6 from 4, 5, 6, and 9, we would obtain the following scores: −2, −1, 0, and 3. (These num-bers will always total zero.) By squaring each value, we obtain 4, 1, 0, and 9. We then add the squares together to obtain the sum of 14. Because there are four scores, we di-vide 14 by 3 ($n - 1$), which gives us 4.66. We take the square root of 4.66 and obtain a standard deviation of 2.16.

Admittedly, it is a heady feeling to wield such a sharp intellectual sword and accomplish so imposing a task as determining the standard deviation of a popula-tion. But it is a task within the reach of each of us so long as we wish it to be so and take the trouble to actually learn the formula. Can the standard deviation help us to discover the origins of life or the fate of the universe? Probably not. It is, first and foremost, a tool for measur-ing the distribution of values in a population. The greater the deviation, the greater the dispersion of these values. We would probably find we could prove this conclusion in a mathematical sense by selecting any four middle-class homes, perhaps from the Tin Cup neigh-borhood in the valley below Snobby Ridge, and calcu-lating the standard deviation for the number of cars owned by those four households. Because there would, in all likelihood, be much less variation in the number of cars owned by the more modest Tin Cup homeowners (generally ranging between one and three automobiles) and because their more humble garages could accom-modate only two cars at best, we would probably obtain a lower value for the standard deviation. Even if the mean value of both the Snobby Ridge and Tin Cup neighborhoods was the same, the standard deviations for the number of cars owned by each would probably be very different. After all, the range of incomes in Snobby Ridge would probably vary more than those incomes in Tin Cup simply because Snobby Ridge would be the more likely home for the superwealthy.

What does this all mean? Aristotle and Plato knew nothing of statistics so we must discard philosophers in favor of mathematicians and mention the fine work of the Russian mathematician Chebyshev who proposed a theorem that he named after himself (Chebyshev's theorem) to clothe the concept of the standard deviation with some relevance. In short, Chebyshev's theorem states that the proportion of any set of values lying within n standard deviations of the mean is always at least $1 - 1/n^2$, where n is not greater than 1.

Rest assured, we are not lapsing into mathematical babble. Those patient readers who have not already tossed this book aside and cynically shaken their heads can believe Chebyshev was onto something very important because he knew as much about the distribution of values in a population as he did about Russian vodka. Suppose we apply Chebyshev's theorem to our Snobby Ridge example and decide we want to find the proportion of the values (number of cars) lying within two standard deviations of the mean. By substituting the number two for the symbol n, we find that at least three-quarters of all the values (the number of cars owned by the Snobby Ridge residents) will fall within the interval ranging from two standard deviations above the mean to two standard deviations below the mean. If we expand to three standard deviations above and below the mean, we will find that nearly nine-tenths of all values will be included within that interval.

What is the practical significance of these intervals? By adding successive standard deviations, fewer and fewer population values will necessarily fall outside the interval. In other words, we incorporate a greater and greater percentage of all of the values in the population and, as a result, obtain a more accurate description of the entire population. When we use three standard deviations both above and below the mean, we are essentially saying that nine-tenths of all the values in the population are contained within that interval. If we go to four standard deviations both above and below the mean, we shall include nearly 94 percent of all the values in the population.

If we were interested in examining the distribution of wealth in this country, Chebyshev's theorem would tell us, if we had obtained the mean value for household wealth in this country, where any single household stood in relation to the population as a whole. After all, we are not so concerned with our wealth in absolute terms but instead with how we stand relative to our friends, neighbors, loved ones, and in-laws. If we determined our household income was more than three standard deviations above the mean but less than four standard deviations above the mean, we would know our income was probably in the top ten percent of the country. Although we would be happy knowing we had managed to climb above nine-tenths of the nation's population in the economic pecking order, we would be sobered by the realization that the top ten percent of the wealthiest households still lay above us. The standard deviation of a population, in conjunction with Chebyshev's formula, can be used to tell us exactly where we stand relative to those of our fellow citizens who were fortunate enough to be born into families with big bank accounts and poor health.

Measures of standard deviation are also important in determining the quality or shoddiness of a product. Manufacturers rely on statistics to help them select representative samples of their products for testing. A tire manufacturer who wished to test the resistance of its tires against nail punctures would not want to test every single tire or else it would be left with nothing more than a pile of punctured tires. Instead the manufacturer will select a tire every so often and subject that tire to testing to determine whether the product is up to its usual standards. The standard deviation can be used to evaluate product quality. If the Missing Finger Tool Company determines that its Mulchmeister has a mean operating life of five years, it will want to determine the standard deviation of its production of Mulchmeisters. If the company finds there is a comparatively large standard deviation, then the corporate heavies will have some cause for concern because relatively large numbers of Mulchmeisters can be expected to fail before they see their fifth anniversary of operating life.

While many manufacturers might secretly desire that their products wear out right away so their patrons will come rushing back to the store to buy a replacement model, this type of business plan does have certain drawbacks. Mulchmeister patrons might be so disgusted with their broken-down Mulchmeisters they might switch to competing products such as Garden Joy's Mulchmoocher or Green Thumb's Mulchmiser. The powers-that-be at Missing Finger will have to focus on their quality control program so that the reliability of the Mulchmeisters can be upgraded and the possible loss in customers avoided. Assuming the corporate hierarchy began to crack the whip on the assembly line and ordered employees to attach moving mechanical parts with nuts and bolts as opposed to string and glue, then the standard deviation would likely decline and the overall product quality rise.

Finding Your Position in the Human Race

One of statistics more charming features is that it can enable anyone to calculate relative positions in a population. Someone (not Einstein) once said, "Everything is relative." Even though this thought may have been prompted more by hallucinogenic drugs than by profound philosophical insight, it does have a certain kernel of truth. We as individuals are concerned about our position relative to our fellow human beings in every part of our daily lives. We are much less concerned with our absolute position (assuming we have the basic necessities) because it is almost impossible to evaluate anything without being able to refer in some way to some type of measurement of comparison.

Money is a good example of something that is more meaningful when we view it in relative (as opposed to absolute) terms. To say I have $100 million is virtually meaningless unless we can examine the value of that sum in terms of the goods it can purchase or the amounts of money possessed by other individuals. To simply have some quantity of any item ("I have 50 rocks" or "I have 700 tables") does not really lead us anywhere. It is simply a statement of possession. If we know $100 million can purchase 5,000 new cars or 250 congressmen or 18 luxury estates or eight office buildings or one Ivy League education (including room and board), then the concept of having money acquires some relevance. It has purchasing power, and the bundle of goods and services it can purchase will change over time due to price fluctuations. Also relevant is the amount of money everyone else has because I can now directly compare my wad of cash with theirs.

Human beings are inherently competitive, forever trying to compare themselves to each other to find out where they stand in the human race. Of course money is a common focal point because many people often become dissatisfied with the amount of money they have once they learn that a person they previously considered their equal (or inferior) is earning more money than they. Even though they may earn more than enough money to pay all their bills and take an occasional trip to

Atlantic City, they will be unhappy because they will perceive their place in the human race as being further behind than they had originally believed. This comparative jealousy extends to the highest strata of society— even billionaires! Indeed, we can imagine a "poor" billionaire who has only a measly $1.2 billion to his name pouting in his Lear jet because he just learned his next-door neighbor located a new oil field and is now worth $1.8 billion. After all, the "poor" billionaire would hardly be able to hold his head up at the polo matches, knowing he had fallen so far behind his fellow billionaires.

Our educational institutions require us to undergo various standardized tests to determine how we compare to each other. It makes little sense to say I received a 4.0 (the numerical equivalent of an "A") in geometry unless we know this is the highest possible score obtainable in a geometry class. We can derive more meaning from this finding if we know that my grade places me at the very top of my class, ahead of my fellow students. The 4.0 is not self-explanatory because it does not mean I correctly answered every question on every quiz and exam. Instead it means I was the top finisher among my classmates. In short, I was the best among an imperfect group. Grading on a curve is thus how teachers rank "varying degrees of imperfection." Although some (usually those at the bottom of the class) would argue that such differences are merely cosmetic, the fact remains that our innate intelligence, analytical abilities, knowledge, and vocational skills are often evaluated using standardized tests. And these tests not only give us grades but also show how we compare relative to the other individuals who took that very same test. For those individuals who are absolutely convinced they are better than everybody else, such tests can prove to be either exhilarating or sobering.

Standardized tests are designed to show how students compare to each other in their mastery of certain subjects such as grammar, algebra, vocabulary, and history. Fortunately for those of us who must know where everyone else stands in society's pecking order, statistics gives us a formula for determining the number of standard deviations by which a given value (such as a student's test score) lies above or below the mean score received by all the students in the sample or population. To obtain this number, we simply subtract the mean score from the individual score and then divide that amount by the number of scores:

$$\frac{x - \overline{x}}{n}$$

If we want to find out how well one student, Bruno, fared in his fashion design course at the university, we would need three pieces of information: Bruno's score (71), the mean value of all the scores in the class (66), and the number of scores (students) that make up the mean (18). By plugging these numbers in the above equation, we obtain the following results: 71 (Bruno's score) − 66 (the class average) divided by 18 (the number of scores) is equal to 5/18 which is equal to 0.27. So Bruno's test is 0.27 standard deviations above the mean of the fashion design class. How could we use this equation to compare Bruno's grades with those of his friends in other classes?

Suppose Bruno and his good friend and fellow student Heinz become involved in a heated argument over who is doing better in their respective classes. Bruno and Heinz do not take any of the same classes, so there is no way they can resolve their dispute by comparing their grades with one another. Moreover, Bruno's studies center around fashion whereas Heinz is majoring in arts and crafts. The difficulty of comparing grades is compounded by the fact the School of Fashion Design grades students on a 100-point scale whereas the School of Arts and Crafts grades students on a ten-point scale.

Bruno and Heinz agree there must be a way to determine who is the better student. They kidnap a statistician named Ed and hold him hostage until he shows them a procedure to compare their test results with each other. Not surprisingly, it involves that very same equation used to show Bruno how he was doing relative to his class. Ed takes it one step further by obtaining similar information from Heinz's arts and crafts class and then comparing the results. Heinz's score is 7, the mean value of all the scores in the class is 5.5, and the number of scores (students) in the class equals 15. By plugging these numbers in the very same equation, we obtain the following results: 7 (Heinz's score) − 5.5 (the class average) divided by 15 (the number of scores) is equal to 1/10, which is equal to 0.1. So Heinz's grade is 0.1 standard deviations above the mean as compared to Bruno's, which is 0.27 standard deviations above the mean. At first blush, Bruno would appear to be doing slightly better relative to his classmates than Heinz. However, the difference between the two is so small as to be statistically insignificant. There is a discrepancy, and Bruno can argue he is slightly better than Heinz in a relative sense. At the very least, they can untie Ed and let him return to his statistical work. Bruno and Heinz can then become embroiled in a fight over who is the better student and who stole who's girlfriend and who called who's mother a very bad name.

College entrance examinations use percentile scores to rank all of the people taking the test. Even though a standardized test may contain a score from 1 to 99, it makes little sense to say Billy Bob received a 54 by itself because we have no idea what the raw score by itself means. The score becomes significant only when we find out how Billy Bob did relative to his fellow applicants and rank all of the test results accordingly. If Billy

Bob learns that he scored in the 54th percentile, then it does not mean Billy Bob answered 54 percent of the questions correctly. Instead it means Billy Bob scored higher than 53 percent of his fellow students and lower than 46 percent of his fellow students. Even though Billy Bob may have "aced" the sections on geometry, vocabulary, and three-fingered milking techniques, he obviously could not have done well in the remaining sections including algebra, history, and hog curing.

We would obtain Billy Bob's percentile score by dividing the number of scores less than Billy Bob's score by the total number of scores and then multiplying that quotient by 100 to convert it to a percentile. This procedure would enable us to convert any score to a percentile and thereby make any relative comparisons desired.

Having become masters at handling a vast range of statistical concepts in this chapter, including such weighty topics as averages, frequency tables, means, and standard deviations, it is now time to turn to a different subject: probability theory. In the next chapter we shall sharpen our card playing and dice throwing ability (only for purposes of self-enlightenment) and also learn about the crucial role played by probability theory in statistics.

Lotteries and Other Improbable Probabilities

In play there are two pleasures for your choosing—
The one is winning, and the other losing.
—LORD BYRON

M ost visitors to Las Vegas who try their hands at any of the many games of chance available for their pleasure—including craps, blackjack, roulette, and poker—have often unsuccessfully grappled with the laws of probability. Similarly, every purchaser of a state lottery ticket is essentially wagering that the laws of probability, however improbable, will somehow see fit to smile upon their single ticket and rain untold riches upon them. Unfortunately, very few of us will experience the exhilaration of winning the world's highest stakes poker game or multistate lottery because the odds are too small.

For those lucky people dealt the royal flush or who pick the correct sequence of numbers, the laws of probability will be warmly welcomed. At least for them, fate smiled and the normally unforgiving laws of chance were warm and beautiful. Such victories will be even more dramatic and sweet because most people play the lottery having little or no real expectation of winning. It is a form of entertainment for them, not an investment. But there are some people who view playing the lottery as a form of retirement planning. The strategy is simple: Hit the lottery and then retire. The problem with this strategy is that there is no guarantee that even the most fervent lottery player will be able to win the big prize. Indeed, the typical state lottery requires the player to pick correctly six numbers on a card. These cards may contain 45 or 50 numbers so the odds of picking the winning combination may be one in ten to twenty million. Nonetheless, millions of people will still go to the counter at the local grocery or convenience store each day and buy a few tickets to try to win a jackpot. Sadly, there are no bonus points or reduced odds for those persons who play every week year after year as compared to the occasional player who just happens to win the lottery on his first try.

Sometimes several weeks or even months pass before someone manages to choose a winning ticket. In the bigger state lotteries, the jackpot is usually rolled over until there is a winner. This means the jackpot may grow to several times its "normal" size. In some of the larger states, it is not totally out of the ordinary for a jackpot that is normally $8 or $10 million to balloon up to $30, $40 or even $50 million before someone (or a few lucky players) manage to select the winning numbers. Of course, this upward surge in the payoff amount will often create a buying frenzy among people who would not otherwise buy lottery tickets. As many new players come into the lottery pool, the greater number of tickets being purchased will usually ensure there will be more winners and hence a smaller total share of the jackpot. Whereas the normal $10 million lottery might have one winner, the lottery in which the prize has been rolled over several times to $40 million might have five, six, seven, or even more winners. The players in the big lotteries might find themselves paradoxically competing for a relatively smaller prize than those "typical" $10 million lotteries with prizes too small to draw the interest of the casual or occasional player.

Events and Experiments

The novice statistician will quickly learn it is nearly impossible to nail down the basic concepts of probability theory without learning to distinguish between "events" and "experiments." An *experiment* is an investigative process that generates observations. If I decide I want to flip a coin a thousand times to see whether it will tend to have approximately equal numbers of heads and tails, then the process of conducting this thousand-toss trial is an experiment. The *event*, by contrast, is the result of the experiment. In the case of the coin tosses, an event would be the result of a single toss—either heads or tails.

Suppose we managed to obtain a grant so that we could hire a professional investigator to determine whether it was equally likely a coin would land on either the "heads" or "tails" side. To express it a little more precisely, the investigator might ask whether it was equally probable that a tossed coin would land on "heads" or "tails." That investigator might design an experiment to test whether either outcome was equally likely by asking whether a coin tossed up in the air 10,000 times would land on heads about 5,000 times and land on tails the other 5,000 times. He would probably pull a coin out of his pocket and begin flipping it. The repeated coin flips would be the experiment and the showing of either heads or tails with each flip of the coin would be an event. The tediousness of this experiment would not lend itself to coverage on the lowliest sports cable television network ("Tonight we will see the finals of the Random Coin Toss!"), but it would be the only way we could test the proposition. By the time the coin had been flipped the requisite 10,000 times, the number of heads and the number of tails could be tabulated and some type of conclusion drawn about the probability of the coin landing to reveal either "heads" or "tails." If either side had an equal probability of turning up, we would expect 10,000 flips of the coin to result in 5,000 heads and 5,000 tails.

What if the coin toss experiment resulted in 4,998 heads and 5,002 tails? Or what if there were 5,006 heads and 4,994 tails? Would these results invalidate our basic presumption that the coin was equally likely to land on either side? No, not really. The fact there was a minor discrepancy between the number of "heads" and the number of "tails" would not discredit our belief either side was an equally likely outcome or event. But if we had 9,900 heads and 100 tails, we would seriously begin to wonder whether there was something askew with our basic assumptions or, more likely, whether there was something wrong with our coin.

If we were to express the relative frequency approximation of this experiment, we could say that the probability of an event A (the coin turning up "heads") occurring is equal to the ratio of the number of times the coin actually showed heads divided by the total number of coin flips. In other words, if we toss the coin 500 times, and 250 of those times turn up heads, then we would calculate the ratio by dividing the 250 tosses that came up heads by the total number of coin tosses. We would express this in a somewhat more mathematical format by saying that P(A), which is a shorthand expression for "the probability of A," is as follows:

$$P(A) = \frac{\text{the number of times "heads" showed up}}{\text{the total number of coin tosses}}$$

So the P(A) when heads shows up 4,998 times out of 10,000 coin tosses would be .4998. The fact that P(A) was not exactly .5000 (which would have been the case if 5,000 heads had shown up out of 10,000 tosses) would not be especially troubling because it would be a statistically insignificant difference (less than one-tenth of a percent). Indeed, we might be more surprised if P(A) was exactly equal to .5000 (or 5,000 heads) after 10,000 coin tosses. After all, there are many things that could occur during those many thousands of coin flips, such as a tidal wave, a civil insurrection, or even a thermonuclear attack.

The Law of Large Numbers

When we use actual tests to obtain probabilities as with the coin flipping experiments, we see our actual results may deviate from the results that would be predicted by probability theory. An interesting thing happens as the number of coin tosses increases: The deviation between our actual results and the results predicted by probability theory begin to narrow. In other words, the more times we flip a coin, the more likely it is that the ratio of "heads" to the total number of coin tosses will more closely approximate 0.5. This means that 1 million flips of the coin will be more likely than 100,000 flips of the coin to result in a perfect 0.5 relative frequency. This tendency of our real world empirical tests to more closely approximate the predictions of probability theory as the number of tests or trials is increased is predicted by the law of large numbers.

The law of large numbers has its genesis in common-sense observations. It basically says it is more likely that a greater number of coin tosses will exactly mirror the predictions of probability theory than will a lesser number of coin tosses. This law was not coined (no pun intended) for numismatic aficionados alone but is applicable to any type of phenomena. Blackjack players, for example, can delight in the comfort of the law of large numbers because it will tell them any streak of losing hands of cards is bound to end so long as they can manage to afford to continue playing long enough. Whether they choose to test the applicability of the law of large numbers by trading in their car title and the deed to their house for chips depends on their confidence that the law will see fit to shine down upon them before they go bankrupt. Of course, the law of large numbers may not be able to save those people who are pathetic blackjack players.

We can take a certain amount of comfort in the law of large numbers because it assures us that things will, at some point, return to normal. If I am a proud father of eight children (all girls) who finds birth control to be an inconvenience, I can rest assured that at some point, the laws of probability are going to swing back in my favor and allow me to sire a male child. Whether my wife wants to continue this statistical experiment five, seven, eight, or even nine more times until the law of large numbers swings into action and gives me a male child is less clear. Even for such things as live births, the law of large numbers tells us we can expect that the births of boys and girls will tend to equalize as more (ahem) trials are conducted. If we decide to press on and continue this test of probability theory by having 31 more children (all of whom are girls), then we may want to reconsider our approach. Certainly my wife will want to consider a divorce, particularly if I insist that the "next time's the charm." If we were to get such an aberrant outcome of 39 daughters and no son, then we may need to examine the entire testing procedure to see whether there is something fundamentally flawed with this experiment. The laws of probability would not predict 39 consecutive female babies. At the same time, it would be difficult to point to anything obvious (in the absence of bizarre fertility rituals or inept procreative techniques) that would readily explain our affinity for daughters. For most people, however, the law of large numbers would constantly reassure them that the number of daughters and sons in a given household would tend to be equal over time.

Into the Casinos

One of the most obvious benefits of studying statistics is that it will help you prepare for the glittering world of casino gambling. Notice we did not say you would win a great deal of money in the glittering world of gambling casinos. Most players are absolutely convinced their chances will be enhanced by rolling a "lucky dice" or becoming involved in a game when another player is having a "lucky streak." Sadly, there is no such thing as luck. A person winning a series of hands in poker or blackjack, for example, has nothing to do with some type of occult force on the loose in the casino. Instead a winning hand merely represents a happy coincidence whereby one player is dealt a better hand of cards than the house or the other players. As such, the win is a testament to the laws of probability because these laws over the long haul are much more predictable and less capricious than the whims of Lady Luck. Indeed, we should take some comfort in the fact that these laws ensure a specific card will be drawn with a certain degree of regularity over time.

Gamblers try to discern recurring patterns in the colors on a roulette wheel or the numbers on a pair of dice or the faces of a deck of cards. This search for "a lucky streak" is one of the greatest mistakes they can make. Remember, all gambling events are independent of each other, and every toss of the dice or turn of the wheel is a completely independent event. Even though the roulette wheel may stop at red for ten consecutive turns, each of those events is independent of each other. The fact that it is extremely unlikely ten consecutive turns of the wheel would turn up red does not violate the laws of probability. Such a series of events may be unlikely, but not impossible. Probability theory does not enable us to predict with certainty when any single event will occur. It simply assures us that each event will occur with greater frequency as more spins of the wheel or rolls of the dice occur.

Because gamblers realize how unlikely it is that a certain color or number will turn up over and over again, they will often bet against the continuation of that series, reasoning that a change in colors or numbers is overdue. As each event is independent, however, there is no statistical advantage in betting for or against the continuation of the streak. Of course the streak will end at some point but no one knows exactly when.

Pity further the player who decides to wager on the occurrence of a specific event. According to Tom Ainslie's *How to Gamble in a Casino*, this faith in overdue occurrences is costlier than faith in prolonged streaks. "The poor soul who bets on 'overdue' events becomes more deeply involved as losses mount and the expected color or number fails to appear."[1] Ainslie's strategy is to focus on skillful betting, which consists of "(1) confining play to the least unfavorable bets; (2) imposing a limit on the amount of money one is willing to lose in each session; (3) imposing a much lower limit on the winnings to be sought at each session; (4) betting systematically; and (5) leaving the table as soon as the winnings or losses reach the prescribed limit. "[2] Ainslie goes on to provide detailed information about the preferred ways to bet in games such as roulette, blackjack, and craps. He repeatedly cautions against the common practice of trying to build small winnings into fortunes or redressing a small loss by continuing to wager good money after bad. He warns that instead of giving into gambling fever, the skillful player should quit when he or she has amassed some funds or has begun accumulating significant losses.

Random Samples

The games of every casino are dependent on the spin of a roulette wheel or the roll of a dice. Casinos have little desire to engage in crooked games because the price of such adverse publicity would far outweigh any possible gain. Needless to say, rumors about crooked gaming tables do little to attract customers. The reputable casinos do not want even the slightest questions about the integrity of their systems to be raised.

Like the casinos, statistics is vitally dependent upon samples that are unbiased and representative of the population as a whole. Selecting a random sample is not as simple as it may first appear. We all have unconscious preferences that affect the types of people or things we include in or exclude from our sample.

Assuming we are able to devise a reliable sampling process, we can utilize our statistical tools to determine the probabilities of certain things or events. If we want to determine the probability that a given kilt-clad Scotsman is not wearing cotton briefs, then we would probably survey 500 or perhaps even 1000 randomly selected Scotsmen and ask them whether they feel extra drafty underneath their kilts. If 630 out of 1000 Scotsmen respond they prefer the tartan wool to cotton briefs, then the probability P(wearing underwear) that a given kilted Scotsman will be wearing underwear is 1000 − 630/1000 or .37. So straphangers on public transportation may wish to avoid brushing up against kilted Scotsmen because there is an almost two-thirds (.63) probability they may see more than they ever thought possible.

Betting on Insurance

Every time we purchase an insurance policy we must deal with probability theory. The viability of the entire life insurance industry is dependent on the actuarial tables developed by insurance companies to estimate the probable life spans of individuals once they reach a certain age. By determining the probability that a certain individual who is 35 years old will reach 50 years of age, for example, the insurance company can then determine the rate it must charge that 35-year-old in order to underwrite a policy for a given amount of coverage. This is not to say these predictions are an exact science. Invariably, some people will die prematurely. The trick for the insurance company is for its actuarial tables to be accurate enough so that, on the whole, the company will collect more premiums from any given age group than it will pay out in claims. Its position is somewhat akin to the "house" in the casino because the "house" always wins in the long term—in a manner of speaking.

Purchasing insurance is somewhat similar to rolling the dice at a crap table. You as the insured are essentially betting that you will die prematurely and thus be unable to provide for your family. Your purchase of the insurance policy will be motivated by a desire to help your spouse and children weather the inevitable financial crises that will follow the death of an income-producing family member. This is not to say the insured individuals are hoping their policies will pay out because they certainly will not be around to enjoy the windfall. Sitting in an urn on the mantlepiece will not enable them to share in the fun when the family members are jumping up and down in delight after receiving a million-dollar check from the insurance company. The purchase of insurance does provide a certain peace of mind for the insured if the unthinkable does happen and he does get run over by an ice cream truck or a meteor bores a hole in his car while he is driving to work.

Do insurance companies like paying claims? No more than anyone likes getting root canal work done without anesthesia. Insurance companies realize there will be a certain number of claims for any given age group because not everybody will outlive the term of coverage of the policy. The major insurance companies will promptly pay even very large claims because that is part of the basic bargain with the insurance consumer. Of course, insurance companies will try to be as accurate as possible when screening their applicants to sift out people who have preexisting conditions (e.g., bad heart, cancer) or unhealthy habits (drug addiction, etc.) so they can reduce the likelihood of having to pay out a given claim. To that end insurance companies require applicants to fill out detailed questionnaires regarding their health and family history of illnesses. This information enables the insurance underwriter to obtain a more accurate picture of the risks involved in insuring the life of the applicant. The issuance of the policy is predicated on the applicant truthfully answering those questions and being subjected to a medical exam. Based on this information, the insurance company can then determine the rate it must charge for a given amount of coverage that, on the whole, will be enough for it to conduct a profitable operation. Should the applicant fail to disclose certain material facts, such as an existing medical condition or terminal illness, at the time of the application process, then the insurance company may refuse to pay the policy if the insured succumbs to that medical condition or illness. The medical examination and questionnaire are used to highlight preexisting medical conditions that may be excluded from coverage.

There are always a few fly-by-night insurance companies that will try to reduce even further the likelihood they will pay out any funds. They will routinely contest any claims filed even when the body of the deceased insured is brought by the beneficiaries to the corporate offices of the insurance company to collect the death benefit. But this is a strategy that has nothing to do with examining the estimated life spans of individuals at various ages; it is merely an attempt to stack the deck further in favor of the insurance company. Fortunately, most established insurance companies will pay legitimate claims in a timely manner even though they would all secretly prefer to insure only people who were destined to outlive their coverage. At least the insurers and those insured have one thing in common: the wish for a long and healthy life. An imaginative insurance company might also decide to unbundle its coverage by insuring against specific diseases so that an applicant might pay, for example, $7 a year for a bubonic plague policy or $2 for a policy insuring against gout. But it seems unlikely this approach would ever find much favor with the public, despite the widespread concern about the cost of insurance.

Probability theory is concerned with the likelihood that a certain event will or will not occur. Statisticians will assign a given probability to an event that will be somewhere in the range between 0 and 1. An event having a probability of 0 is an impossible event and cannot occur. By comparison, an event having a probability of 1 is certain to occur. An event that has some probability, however small, of occurring will fall somewhere between those two extremes and will, hence, have a numerical value between 0 and 1. By way of illustration, the probability that the movie rights for this book will be purchased is very unlikely. It is not equal to 0 because we cannot say for certain that some studio executive will not one day decide that it was high time to film a motion picture about statistics for the millions who have dreamed of such an entertaining film—good, wholesome family entertainment. The likelihood of such an event occurring is fairly remote. In contrast, the probability this book will make a profit for its publisher is fairly high (perhaps .6 or .7), but it is not a certainty unless the author commits to purchasing several thousand copies with his own funds. As far as events at either end of our probability continuum are concerned, we can say that it is certain we will all eventually die. So the numerical value we would assign to this event is 1 because that is the number that represents the sum total of all the possible events in a probability distribution. Yet we can also be equally certain there is no likelihood any human being will be able to leap off a building and fly through the air without the assistance of rocket belts or some other mechanical devices.

Multiple Events and the Addition Rule

Up to this point, our discussion of probability theory has dealt only with single event experiments. In the real world, however, many events actually are composites of two or more simple events. In those situations, the statistician must determine the probability that one or more events will occur during a particular experiment. For example, a company interested in determining the rate of defects in a given batch of five computer monitors may want to calculate the probability that the first four monitors will work properly. As such, it will have to calculate the probabilities of selecting a working monitor for four consecutive times and not selecting the defective monitor in that batch.

    We may use the notation P(A or B) to represent the probability that event A will occur or event B will occur *or* that both events A and B occur together. If we add the number of ways event A and event B can occur together, we must be careful to avoid counting twice those outcomes in which both A and B occur together.

Suppose a professor of chemistry formulates a chemical that will grow hair on the palms of a person's hands. His excitement about his expected financial windfall coupled with the worldwide fame that would result from finding a true cure for baldness would probably prompt him to round up a few students to test out his product. If he gave 30 students $5 each to try the tonic, gave another 30 students $5 to drink plain tap water, and then gave nothing to the final 40 students (the control group), then the probability of randomly selecting a student who had either consumed the tonic or was in the control group would be P(tonic or control) = P(tonic) + P(control) = .30 + .40 = .70 (which would represent 70 percent of the 100 students). Suppose further that 20 of the students who consumed the tonic showed appreciable new hair growth (albeit on their backs or the roofs of their mouths), and six who consumed water also displayed new hair growth. The probability of randomly selecting a student who had either consumed the tonic or who had grown hair would *not* be equal to 30 + 26 (the number of students using the tonic + the number of students who had grown hair). This total of 56 would involve some double counting because many of the students who used the tonic were also in the group that had grown hair. The probability of selecting a student who had either consumed the tonic or grown hair would be equal to 30 + 6 (those non-tonic consumers who had still managed to have new hair growth occur) or 36. In a group of 100 participants, we would thus have a probability of .36 because we would have taken care to avoid counting some of the students twice. We might express this procedure for avoiding the double counting in the following way: P(tonic or new hair growth) = P(tonic) + P(new hair growth) − P(tonic and new hair growth), which will give an accurate calculation of the probability of selecting an individual who had either consumed the tonic or grown hair.

This addition rule may be expressed algebraically as P(A or B) = P(A) + P(B) − P(A and B). Statisticians usually replace the expression "or" with a sort of upper case "U" and the "and" with an upside down "U" to be concise. Of course the concern with double counting disappears if events A and B do not occur simultaneously. In such cases, we say that the events are mutually exclusive of each other because it is impossible for them to occur together.

If we go to a Las Vegas casino and play a hand of blackjack, the probability of being dealt any given card in a deck is 1/52 or .019. Once we have been dealt a particular card, such as the ace of spades, then it is impossible for us to be again dealt that very same card using that deck. If we receive three or four jokers in the course of a game using a single deck, then we might rightfully ask the dealer: "What's going on here?" or some other pointed question designed to get to the bottom of this mess. If we were consistently winning with a "blackjack" hand, however, we might want to keep our mouth shut. Yet, if the dealer was using several decks of cards for a single game, we might not have such cause for concern.

Two events are said to be mutually exclusive if they cannot occur at the same time. If we are dealt an ace of spades, then we cannot, by definition be dealt a three of clubs or a four of hearts. The selection of the ace of spades automatically precludes any other card from being dealt at that very moment. The fact that a five of diamonds could be stuck to the ace of spades would not somehow vary our point because we are allowing only one card to be drawn at a time.

Statisticians must also deal with events that are not mutually exclusive in that they can both occur at the same time. If we were conducting a survey of strippers, we could select a stripper who moonlighted as a pale-ontologist and was also under 30 years of age because these two events would not be mutually exclusive. Similarly, we could select a stripper who was schooled in the modern theories of macroeconomics and was also a natural redhead because one can both study the works of John Maynard Keynes and have a lovely auburn hairstyle.

What is the point of this discussion of mutual exclusivity? It brings us back to our earlier discussion of the addition rule in which probabilities for pairs of events can be calculated. This rule does force us to evaluate whether these events can occur simultaneously and, if so, whether we have avoided the double counting that is invariably present in non-mutually exclusive events. Of course a pair of mutually exclusive events poses no such problem. There is no possibility of double counting the events using the addition rule because they cannot occur simultaneously. We cannot, for example, select a person who has type A blood and type O blood because a person has only type A or type O or one of the other blood types. Selecting a person with type A blood is an event that is mutually independent of selecting a person with type O. In other words, it is impossible to find both types of blood naturally present in one person—at least in the absence of a transfusion. A non-mutually exclusive event might involve selecting people based on hair color, particularly if some people tended to have many different colors of hair. As a result, we could select a person with blonde hair, a person with black hair, or even a person with both types of hair—albeit chemically enhanced.

Multiple Events and the Multiplication Rule

Although we have already learned much about the calculation of probabilities that either of two events will occur in a given experiment, we have not up until now considered a way to calculate the probability that both of these events will occur at the same time. While the addition rule was fine up to now, we must turn to more sophisticated weapons to humble those who would dare suggest we might have better things to do with our time than study statistics.

What makes the multiplication rule particularly interesting is that it enables us to calculate the probabilities that both events A and B will occur. We can get some idea of how the multiplication rule works by considering the social engagements of the retired anthropologist, Professor Miles Hester. Our Professor Hester has always been involved in torrid romances with lovely, but headstrong, young ladies each semester. Dating several young women at a time, he seems to take an almost bizarre delight in running the risk he will be caught in a compromising position by one of his other paramours. On a given Wednesday night, Professor Hester is planning to go to a local restaurant for dinner and drinks with Helga, an unlicensed physical therapist to whom the professor has graciously offered his body to use as she sees fit. He then plans to end his date with Helga and meet the studious but smoldering Dr. Judy Forman, a noted veterinarian, for an evening of dancing at the Spiced Sow Cafe. What are the odds Professor Hester will actually be able to keep his dates with both women without getting caught?

Sadly, statisticians have not been able to use the multiplication rule to calculate the probability for avoiding such a calamity or such an exhilarating event—depending on one's viewpoint and betting strategies. We can illustrate the way in which the multiplication rule works if we assign arbitrary probabilities to both events. If Helga is in great demand due to a recent influx of German tourists, then the probability she will be available to share a dinner with Professor Hester may be somewhat low, such as 1 in 5 or .20. On the other hand, Dr. Judy Forman may still have the lingering scent of smeared cat intestines in her hair and thus not be in great demand as a date. As a result, there will be a reasonably high probability she will be available to meet with Professor Hester for a magical evening of clogging—perhaps as high as 3 in 4 or .75.

If we know the probability that Helga will meet with Hester for dinner is .20 (P(Helga) = .20) and the probability that Judy will meet with Hester for dancing is .75 (P(Judy) = .75), then the multiplication rule tells us the probability that Hester will meet with both Helga and Judy in the same evening is P(Helga) × P(Judy) = .20 × .75 = .15. So the odds may not be as promising as Hester might like to believe that he will be able to keep his dates with both ladies.

Our calculation of the probability that Hester will be able to keep both dates depends on both events being independent of each other. In other words, the occurrence or nonoccurrence of the date with Helga has no bearing on whether Hester will meet Judy for a bout of groping on the dance floor. This result might best be assured if Helga and Judy lived in different countries and never set foot in the same cities. But if Hester is to have a proper social life, transcontinental dating practices cannot be taken on lightly because of the tremendous amounts of time and energy required to sustain such relationships. Therefore, Hester must run the risk that Helga and Judy could meet and compare notes as to their recent experiences with men and find out they were dating the same Hester. Presumably, neither woman would be very excited about the prospect of sharing Hester with the other. The fact of the matter is that if Judy were to find out about Hester's earlier evening with Helga, then she, despite the lingering stench of cat intestines, would be less likely to be thrilled with the prospect of dating him and might not bother to show up at their prearranged meeting place. Judy's knowledge of Hester's date with Helga would compromise the independence of the two events—the dates with Helga and Judy—because Judy's decision to date Hester would be colored by her knowledge he had already seen Helga.

In terms of our multiplication rule, we would no longer be able to simply multiply the probability of the first event (the date with Helga) by the probability of the second event (the date with Judy) because the date with Judy is a dependent event. It will necessarily be affected by the fact that Hester first dates Helga. A statistician would express it in a more precise way, offering a multiplication rule for dependent events. The probability of both events A and B occurring (the dates with both women) would be equal to the probability of event A multiplied by the probability of event B *given event A*. In other words, the statistician would attempt to calculate the likelihood that the date with Judy would still occur given the fact the date with Helga occurred earlier in the evening. Alternatively stated, P(Helga and Judy) = P(Helga) × P(Judy | Helga). Once again, statistics has provided us with a wonderful illustration of how it can be used to wade into the most vexing of problems— even the perils associated with dating more than one person simultaneously in the same evening.

This notion of dependent probabilities is not new but was first proposed by Thomas Bayes, who formulated what is now known as Bayes' theorem. Bayes' theorem is not widely known among mud wrestlers, but it is critically important to statistics because it allows us to revise our probability calculations as we obtain additional information about an event. Indeed, we saw how this theorem can come into play in Hester's dating situation once Judy learned of Hester's date with Helga. Bayes' theorem has a wide variety of applications because it provides a general framework for calculating the probabilities of dependent events of nearly any degree of complexity.

Bayes' brilliant insight about the probabilities of events changing as we learn additional information about a given event is sometimes misapplied, particularly by those who subscribe to services that purport to offer lottery numbers that are either "long overdue" or "hot." The underlying presumption of these publications is that the likelihood of drawing a particular number over time will vary because it has already been drawn many times or has not been drawn in an extraordinarily long time. The flaw in this reasoning is that lottery numbers are randomly drawn. In other words, the selection of each lottery number is an independent event in a given weekly drawing.

Even though the number "10" may be drawn two weeks in a row, there is nothing in the laws of probability to preclude it from being drawn again the next week or not being drawn for another year. These numbers are randomly selected and, as a result, are independent of each other. So while I might be convinced that the first six odd numbers (1, 3, 5, 7, 9, 11) are "hot" and the key to riches as they were drawn in that sequence the previous week, there is no mathematical basis for my believing those particular numbers have an above average chance of being selected because they are independent events. I should keep in mind, however, that my selection of a common pattern of lottery numbers, such as ascending series of even or odd numbers or the first six numbers on the card, may have a drawback because my winnings will be reduced accordingly by all the other winners who chose that very same pattern. By the same token, we need to keep in mind that every selected number goes back in the pot each week; it is not used up for some indeterminate period of time simply because it was drawn as part of the winning lottery number. Even when a player is rolling dice at a crap table and the dice continue to turn up a particular number, the fact this very same number shows up seven times does not make it any more or less likely it will show up an eighth time. Of course, this assumes we are playing with ordinary dice and not a pair that is coin-shaped with the desired number stamped on both sides.

Similarly, we can use our knowledge about independent events to assist in our business affairs. Suppose I am the fleet purchaser for the Billingsly Rental Car Company, where "Quality Is Cool." Further suppose I am trying to acquire ten defect-free Toady automobiles from their East European manufacturer, Sofia Motors, because they are extremely inexpensive. Despite their brick bumpers and their hand-cranked fuel pumps, the Toady is supposed to have very few mechanical problems. Indeed, Sofia claims that nine out of every ten Toadys has no defects at all. As the Billingsly fleet purchaser, I would want to calculate the probability I could purchase ten of these cars without getting a lemon. Accordingly, P(10 defect-free Toadys) = $0.9 \times 0.9 \times 0.9 \ldots 0.9$ (10 factors) = .348. Whether this probability would be high enough to inspire me to complete the purchase of Toadys would, of course, depend upon my own assessment of the quality of the cars themselves.

The use of the multiplication rule and Bayes' theorem can also be of invaluable use at cocktail parties. Suppose you are attending a wine-tasting party where you are allowed to sample any three of the nine unmarked bottles at your table so you can give an unbiased evaluation of each wine. The only information you are given beforehand is that three of the bottles are from France, three from Mexico, and three from Barbados. Not being particularly interested in Mexican or Caribbean wines, you decide you must calculate the probability of selecting the three French wines as your choices. We are not dealing with an independent event here, however. Each selection will reduce the total number of remaining bottles and thereby alter the probabilities of selecting the remaining desired bottles. If we define the probability of getting the three bottles of French wine as events A, B, and C, respectively, then we can calculate the probability of selecting the three French wines as follows: P(A and B and C) = P(A) × P(B | A) × P(C | A and B) = 3/9 × 2/8 × 1/7 = 6/504. In essence, we are saying we are trying to calculate the probability of selecting bottle A and then selecting bottle B, given the fact we have already selected bottle A, and selecting bottle C, given the fact we have already selected bottles A and B. The likelihood I would be able to pull three French bottles out of the group would not be much better than 1 out of 100. As with the Toadys, my assessment of whether to proceed would depend, in no small part, on my evaluation of the likelihood I might suffer some horrible gastrointestinal malady if I were to drink from the wrong bottle.

Complementary Events

Statisticians have also added complementary events to their bag of tricks to help them calculate the probabilities for certain events that would otherwise be very difficult to determine. If the probability of an event A is .1, then the probability that not A will occur will be .9 because not A is the complement of event A. When the probabilities of an event and its complement are added together, they equal 1. In other words, the event and its complementary event are mutually exclusive because one or the other must, by definition, occur.

No doubt, most people would be very impressed by our steel trap logic but they would still wonder why we are bothering to define something that appears to be so obvious a concept as complementary events. The primary reason is that this concept can be used to solve certain types of problems that would otherwise take a lot of time and consume reams of paper. To return to our gambling example, we might want to know, if we were standing at the craps table in a Las Vegas casino, the odds of rolling a particular number such as 5 in the next five rolls. As a result, we want to find the P(rolling 5 in the ensuing five rolls of the dice). As we are playing with a six-sided dice, we would simplify our task by first calculating the probability of P(not 5) for five successive rolls of the dice: P(not 5) = $5/6 \times 5/6 \times 5/6 \times 5/6 \times 5/6$ = 3125/7776. As we have decided, after talking with our personal phone psychic, to pursue the 5 strategy with gusto, we would be pleased to know that the P(rolling 5 in the next five rolls of the dice) = 1 − 3125/7776 or 4651/7776 or almost 0.6. This probability does not guarantee we will actually roll a "5," but it does illustrate how the probability of rolling a given number will increase with successive turns. The one fly in the ointment is that cards and dice have no memory and, like most inanimate objects, know very little about probability theory. Even though the odds for rolling a 5 will increase with each successive roll, there is no guarantee we will not be standing at the table for the next 30 years trying to roll a 5. Of course the odds that a 5 would be so long in forthcoming are laughably small but the fact that each roll of the dice is an independent event does not completely preclude such a farfetched outcome.

Even though we all would like to win a lottery or hit the jackpot at a Las Vegas or Atlantic City casino, we need to bear in mind that probability—not luck—is the paramount consideration. The notion that one can, or even should, engage in a mind-numbing evening-long battle at the blackjack table is at odds with the prescriptions for successful gambling given by the experts. Indeed, the quickest way to lose money and to ensure the continued financial integrity of the casinos is to try to guess when a "hot streak" will begin or end or to somehow search in vain and at great personal expense for a "lucky number." So patience and steadfastness may serve the gambler as well as it did the Tortoise when it competed in a race with the Hare.

From Laboratory Experiments to Casino Glory

Defendit numerus: There is safety in numbers.
—Anonymous

L ike all branches of mathematics, statistics is sometimes criticized for being overly abstract. Our discussion to this point illustrates the fallacy of this view
because statistics clearly has many real world applications ranging from sampling populations to calculating
the rate of defective components in a production line. Its
mathematical basis also gives it another advantage in
that statistical theory provides a pristine model of how
any of a set of events should occur over a given range of
probabilities. Even though the events we observe in the
real world—whether flipping a coin or tossing dice or
spinning a roulette wheel—do not often occur exactly as
would be predicted by probability theory, the theory is
valuable because it provides us with a sort of road map
as to what we should expect would occur if the real
world was as perfect as that of the mathematical world.

The ease with which statistics can be used to model
reality serves an additional function in providing a

check on the events we observe in the real world. Suppose you are playing a high-stakes game of blackjack in a casino and the dealer wins ten consecutive hands with scores of exactly 21 points in each hand. You, on the other hand, have been unable to come within five points of a 21-point hand, let alone "blackjack." If you were the cynical, mistrusting type, you might have a suspicion the dealer was up to no good. If you were versed in probability theory, you could pull out your trusty calculator between hands and calculate the probability of the dealer drawing so many winning hands. You might also look for subtle yet convincing signs of cheating by the dealer, such as the fact your cards have faces on both the front and back. Absent such an obvious sign of underhandedness, however, statistics would provide you with a clue that something was amiss. You could use statistics to calculate the probability the dealer could draw ten consecutive "21" hands. Given the fact such a probability would be extremely remote, your suspicion the dealer was a scoundrel (who showed great promise for a career in politics) would certainly be reasonable. The underlying point that should not be ignored, however, is statistics, by its ability to provide precise probability distributions, enables us to determine fairly easily whether we are witnessing a series of events that are truly random as opposed to premeditated. Of course, it could be that our dealer did, without any trickery, draw the ten "21" hands, but the remote likelihood of the same would probably cause us to disregard such a conclusion.

Random Variables

When statisticians conduct experiments that generate observable events, they find some events are quantitative and others are qualitative. A *quantitative event* might be a measurement, whereas a *qualitative event* might be a subjective evaluation, such as the rating of a figure skater's performance. But because figure skaters are not content to all finish in a tie for first, it became clear very early on that some type of numerical scale would have to be adopted to rank the performances of the skaters. As a result, we see that the judges will assign numerical values to a given skater's performance—even though the basis for such an assessment is admittedly completely subjective. We have seen that even purely subjective judgments can have common standards. Certainly one cannot fail to notice the extent to which the scores given by judges of a figure skater often do not vary that much from each other—except, possibly, when the judge is from the same country as the skater and on the payroll of that nation's athletic team. Of course, every figure skating competition has its share of absurd scores, but the bulk of the scores tend to be fairly close together.

Random variables are the numerical values statisticians use to represent the outcomes of a given experiment. A random variable can take on any type of value depending on the way the experiment is defined. If we decide, for example, to give a class of language students a 200-question surprise quiz in advanced calculus just to see them squirm uncomfortably in their seats, we might define the random variable as the number of answers correctly answered by the students. That variable could have any value between 0 and 200. We also need to recall that random variables may be discrete or continuous depending on the type of operation that is being carried out. In the case of counting the correct answers on the advanced calculus test, the random variable would be discrete because it can only have a limited number of distinct values. In other words, a language student might get 1, 2, or 5 questions correct, but the language student could not get 6.255589 questions correct because there is no such fractional random variable. One can only get a whole number score. Of course there are some teachers who try to muck up the process by giving partial credit for incorrect problems. But for all practical purposes, the scores will be calculated using only whole numbers. So the number of possible values are countable and they are necessarily finite.

Continuous random variables, by contrast, have infinitely many values. There are no distinct intervals. Instead these values spread over an entire range of possible values. Continuous random variables are associated with measurement operations such as heights, weights, and lengths in which there can be infinitely many measurements, depending on the degree of precision that the measurer wishes to utilize in his experiments. We could attempt to make the world's most precise ruler by marking off each inch, each half-inch, each quarter-inch, each eighth-inch, and so forth. We could continue this process indefinitely—gleefully marking every sixteenth-inch, every thirty-second-inch, and so on, until our ruler could measure any given length to any desired degree of precision. But, the marks would soon become so close together our ruler would essentially be nothing more than a solid black line. Yet there could be an infinite number of separate values that could theoretically be measured between the endpoints of this ruler even though it would obviously be impossible to place an infinite number of marks on the edge of the ruler. Of course there are those resoundingly hardheaded types who would simply insist that a sharper marking pencil would do the trick. Even the finest point cannot mark an infinite number of lines. So continuous random variables may seem somewhat mysterious at first because they are not discrete and countable. Instead we must trust that these variables do include an infinite number of values within their boundaries. As we are all busy people, we cannot take the time to deal with an infinite range of values. We typically will focus only on certain interval values such as the half-, quarter-, and eighth-inch marks commonly found on the edge of most rulers. So the continuous random variable will begin to look somewhat similar to discrete random variables because we have no choice but to limit our inquiry to a limited number of values. After all, most people would have very little use for a ruler that measured lengths to the nearest 0.0000000001 inch because they would not even be able to see the spacings between the markings so as to be able to utilize such a wonderful tool. Of course, we do have scientific instruments that can measure very small distances but we do not use ordinary rulers for such tasks; their utility is limited to measurements relevant to our own real world perspective.

Probability Distributions

Statistics is fascinating to so many people because it helps us to make predictions about the frequency certain events will take place in a given experiment. Every gambler is acutely aware there are certain probabilities that a given card will be drawn from a deck or that a given roll of the dice will yield a specific number. Unfortunately, the real world is somewhat messy and does not always follow the pristine predictions of statistical theory. This divergence between theory and fact causes many gamblers to suffer anxious moments and, occasionally, broken kneecaps. The incredible power inherent in probability theory becomes evident over time when we find that as the number of trials in an experiment increases—whether it be the drawing of a card from a deck or the rolling of dice on a table—this divergence between theory and reality will narrow to almost nothing. If we were to hold infinitely many such trials, we would expect this divergence would disappear altogether. Unfortunately, there are very few paying jobs that will allow someone to play cards for the rest of his or her life in order to test the validity of probability theory—except for the house dealer. Consequently, we must devise tests that will confirm or falsify the basic principles of this theory. In addition, we can use histograms to show pictorially the relative probabilities of each value of a random variable where the vertical axis of the diagram will include the probabilities themselves and the horizontal axis will consist of all the random variables. The actual graph as shown in Figure 3 will consist of shaded areas that correspond with the probability of each numerical outcome so that we will be able to determine at a glance the relative probabilities of various

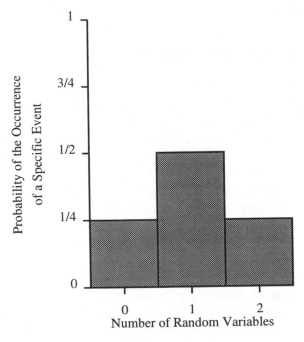

Figure 3. Probability histogram.

random variables simply by comparing the relative areas of the rectangles over each random variable.

We can better explain the gobblygook of the probability distribution concept by offering an example. Suppose the famed university scientist Dr. Olaf Olsen invents a pill that causes any person to grow feathers on top of his or her head. Dr. Olsen believes this "plumage enhancer" is a more stylish cure for baldness than existing hair regeneration formulas because a consumer can grow any type of plumage in colors ranging from blue to green to red to black. Accordingly, Dr. Olsen begins testing the pill on a variety of balding graduate students.

Let us further suppose that in one test he gives the pill to two "plumage-challenged" individuals to see whether both, one, or neither of the individuals will be able to grow plumage. In this experiment, there will be three random variable values (0, 1, 2) to indicate whether neither, one, or both of the students actually grow beautiful feathery crowns. These values will be marked along the horizontal axis of our graph. The probabilities of each of these outcomes will be marked along the vertical axis. Dr. Olsen has determined there is a 1/4 probability that neither subject will grow plumage, a 1/2 probability that only one of the subjects will grow plumage, and a 1/4 probability that both subjects will grow plumage. Although Dr. Olsen has not calculated the probability his medical license will be yanked when the state licensing board finds out about his experiments, he can graph the probability distributions for this experiment. The probability histogram would look like a bar chart as shown in Figure 3: The first bar over the 0 value for the number of cures on the horizontal axis would rise to the 1/4 probability mark on the vertical axis. The second bar over the 1 value for the number of cures on the horizontal axis would rise to the 1/2 probability mark on the vertical axis. Finally, the third bar over the 2 value for the number of cures on the horizontal axis would rise to the 1/4 probability mark on the vertical axis.

While you would agree this is a very fascinating discussion about the various probabilities that neither, one, or both of the subjects will be able to experience the joy of having plumage on top of his head, this probability distribution is providing us with some very interesting information. More specifically, it is suggesting that some experimental outcomes are more likely than others and that as we move to the most common event (one of two subjects being able to grow plumage) the probability reaches a maximum. If we graph the outcome of this experimental trial, we will see the probability function begins at 1/4 (0 subjects grow plumage), rises to 1/2 (one subject grows plumage), and then falls back to 1/4 (2 subjects grow plumage). So the probability histogram will automatically tell us which outcomes are most likely to occur. As with most probability distributions, this experiment will represent a sort of very primitive "bell curve" (which we shall discuss in some detail later). For our purposes, we should simply realize most probability distributions will congregate about some middle value or values and that the outlying or extreme values will, in most cases, be less likely.

There are two very fundamental points that need to be kept in mind when dealing with any probability distribution. First, the probability of any given value of x [the random variable as designated by $P(x)$] in a probability distribution will be equal to or lie somewhere in between the values of 0 (no probability of occurring) and 1 (completely certain probability of occurring). In Dr. Olsen's plumage experiment, we saw that each of the probabilities lay somewhere between 0 and 1: 1/4, 1/2, and 1/4. Second, the sum of all of the individual probabilities must equal 1 so that all possible values are included. Because the probabilities $1/4 + 1/2 + 1/4 = 1$ in the plumage experiment, we know that Dr. Olsen has also satisfied this requirement. As a result, we can whistle a happy tune even as Dr. Olsen's laboratory is being raided by federal agents.

Means, Variances, and Expected Values

We saw previously that a great deal of useful information can be gleaned from a collection of data by determining such things as the mean or average and the dispersion of the values contained in that data. This is also true when one is dealing with the data contained in a probability distribution. Suppose you are sitting alone at home one evening and you decide (perhaps after having had a bottle of wine) that it would be a lark to determine the mean and variance for the number of "heads" you would get from flipping a coin two times. Being a master of statistical theory, you know there are several different outcomes that can take place during these two trials. More specifically, you know the probability that the coin will turn on either "heads" or "tails" is 1/2. The same is true if the coin is flipped a second time. Because of the fact there are only two possible outcomes for the two trials (coin tosses), we know that the probability you will obtain two consecutive "heads" is 1/4. Each event (each "heads") has a probability of 1/2. You would therefore multiply 1/2 by 1/2 to obtain the probability of flipping two consecutive "heads," which would be equal to 1/4. The same calculations would hold true for determining the probabilities of obtaining two consecutive "tails," which would also be equal to 1/4. So the probability we will toss two consecutive "heads" or two consecutive "tails" is 1/4 + 1/4 or 1/2. As you will no doubt surmise, the probability we will toss one "head" and one "tail" (regardless of the order) during the two tosses is 1/2 [Note: The probability of tossing one "head" followed by one "tail" is 1/4 and the probability of tossing one "tail" followed by one "head" is 1/4 so the summed probability of tossing one head and one tail in either order is 1/2.] Thus the probabilities of all of these different outcomes will add up to 1 because these are all of the possible outcomes that can be obtained from flipping a coin twice.

Although thinking such profound thoughts would leave most mortals exhausted, we shall now push forward to incorporate the concept of frequency into our discussion of coin-tossing probabilities. To determine the mean value for the probability distribution, we would add the total number of scores ("heads") and then divide that amount by the number of trials. If you carried out 40 trials (each trial consisting of two coin tosses) and obtained 10 trials with two "heads" and ten trials with two "tails" and 20 trials with one "head" and one "tail" (or vice versa), then you would obtain the mean value for the number of heads as follows:

$$[(2 \times 10) + (1 \times 20) + (0 \times 10)] / 40 = 1$$

So the mean value for the number of "heads" that will be obtained over the course of these trials is 1. The frequency distribution is shown in the equation in that ten trials will yield two "heads," 20 trials will yield one "head" and ten trials will yield no "heads."

Armed with this information, you can then swoop down and determine the variance for this probability distribution. To return to our coin-tossing example, we would take the mean value, which is 1. We would then subtract this value from each of the individual values of the population and then square the differences. We would then add up all the squared results and divide this total by the number of trials. This calculation would enable us to obtain the variance. If we took the square root of this number, we would obtain the standard deviation.

This is not to say the mean will necessarily be a value that can even occur in a single trial. In other words, we might carry out our coin-tossing experiment for a certain number of trials and obtain a mean value that is not an integer (e.g., 1, 2). But the mean does provide us with some clues as to how the values of the aggregated data are weighted. The standard deviation provides us with additional insight as to how the values are clustered around the mean. If the standard deviation is relatively large, then the values are dispersed. If the standard deviation is small, however, then there is less dispersion and the values are clustered about the mean.

Closely related to the concept of the mean is that of expected value. The *expected value* is essentially the average we would expect to obtain if we were to carry out a given experiment for an infinite number of trials. It is particularly important to those of us who try to make our living playing numbers because it illustrates the likelihood that we will lose our money over the long term. Mario Triola's book *Elementary Statistics* provides the following illustration of the concept of expected value.[1] Suppose we play a "three-pick" game in which we select any three digit number and pay $1 for the privilege of picking that number. Further suppose our ticket, like every other ticket in this game, will have 1 chance in 1,000 of winning. How much will we receive if we win? If the "house" had no advantage, we would receive $1,000. But these games are generally skewed in favor of the house so that it can guarantee it will, over the long term, come out ahead. As a result, a winning ticket might only pay $500. While the probability of winning is only 1/1000 with any given ticket, the payoff is not large enough to enable the winning ticketholder to stay even with the house over time. On average, the purchaser of such a ticket will find the expected value for a given $1 bet is 50 cents. So our gambler can expect to lose 50 cents for every dollar bet he places because he has only a 1/1000 chance of winning a $500 jackpot.

This is not to say everyone is destined to have this result because some people will win and quit playing while they still have a profit. It is clear that numbers betting is not an activity for hardheaded investors because it is inevitable it will be a money-losing activity over time. The extent of the disadvantage faced by the bettor will also necessarily vary with the establishment. The advantage enjoyed by the house in this numbers game is typically far greater than that enjoyed by a Las Vegas casino. Moreover, there are significant variations in the expected values associated with games such as slots, roulette, and craps. Not surprisingly, the slots live up to their names as "one-armed bandits" because they do not typically offer odds that are as favorable as craps, roulette, and blackjack. According to Jerry L. Patterson's *Guide to Casino Gambling*, slot machines generally have an 83 percent minimum return rate, which means the casino over time will keep 17 cents out of every dollar played.[2] Some machines offer more favorable odds, but none will provide the gambler with a great deal of assurance that untold riches will be won. Patterson points out a reason slot machine players may have no real minimum return rate: "Almost all slot players lose a far greater portion of their bankroll than the 17 percent edge held by the house because they add their winnings to their bank and keep right on playing until all or almost all of it is gone."[3] Patterson offers this advice for slot players: "Buy as many rolls of coins as you are willing to risk, and as you play, put all the winnings in your pocket. When you lose the last coin in the last roll, quit!"[4] This cautious approach will ensure over the longer term that you will at least be able to take home the minimum return rate and will help you to prolong your gambling experience.

The only way you could expect to break even while gambling over time would be if you were betting at a casino where the expected payoff and the odds were exactly equal. In other words, the casino would have slot machines in which every dollar played was ultimately returned to the players in a jackpot payoff. Unfortunately, such an establishment would soon have to close its doors because it would make no money. Casinos are dependent on making money, so you should not hold your breath that a completely "fair" casino will open up anytime soon.

Binomial Functions

Statisticians who hang around the "beautiful people" know there are certain types of discrete probability distributions functions that have a sort of dual element in that something either succeeds or fails, or something either works or malfunctions. A critical part in a jet engine, for example, may fail and cause the plane to plummet to the earth. Similarly, a new vaccine will either cure its patients of their compulsion to belt out opera arias while sitting in movie theaters or it will not. In such situations, we are concerned with whether a specific part or thing works or not. Thus it makes little sense to talk about degrees of functionality because the object is of concern only if it works or fails at a crucial time. We would not speak of a component in a jet engine as "sort of functioning" nor would we say we had "sort of a problem" to the passengers on that jet as it began to plummet to earth. These situations give rise to a certain type of probability distribution known as a *binomial distribution.*

A binomial distribution has nothing to do with sexual orientation, although its wide range of applications in industry would cause any quality control manager to experience fits of ecstasy. A binomial distribution is a discrete probability distribution that has a limited number of independent trials. The probabilities for each of the trials do not change, and all outcomes of all trials must be classified in one of two categories. The notion of "pass" and "fail"—which is of such crucial importance to manufacturers of any type of product or component part—is the most obvious arena in which the binomial probability distribution is used.

How does the binomial distribution work? Well, if we have an experiment that meets the criteria spelled out in the previous paragraph, then we can determine all of the relevant variables, including the number of trials, number of desired successes, and the probabilities of failure and success in those trials. We can either pull out our trusty calculator and plug these values into the binomial probability formula (a must for masochists) or we can look in the back of any elementary statistics book and find the relevant binomial probability in a table created for just that purpose. While it would be somewhat laborious to plug a bunch of variables into the binomial probability formula, we shall take the easy road and resort to the table.

To illustrate the way the binomial distribution can assist the titans of industry, we should illustrate its usefulness by highlighting a common situation faced every day by factories throughout the world: quality control. The problem begins with the fact that no manufacturer can make a perfect product every time. The Zippy Cola Company is no different. Sometimes its soda cans are so highly pressurized they lay waste to entire neighborhoods when they are dropped. Other times, the syrup content is so high the soda itself has the viscosity of molasses. Still other times, the pull tab edges are so sharp that more than one customer may find a middle finger laying on the floor after opening a can. As a result, the Zippy Cola Company is acutely aware that it must be able to determine the probability a certain minimal number of cans in each twelve-can "Zip Pak" will be without defects.

Suppose the Great Boss of the Zippy Cola Company determines, based on his prior experience, there is a 10 percent defect rate in its products. Let us further suppose the Great Boss wants to know the probability that exactly 10 cans in the 12-can "Zip Pak" will be acceptable. Because the Great Boss is a very busy person (what with golf every afternoon) and cannot have his employees lolly-gagging around (due to the massive employee layoffs the previous year that were necessitated by the Great Boss's $12 million bonus), he would doubtless applaud the use of a binomial probabilities table to answer his question.

Wilfred Sims, chief of quality control, would set up this problem in the following way: Because there are twelve cans in the "Zip Pak," there would be a total of 12 trials or experiments. As the Great Boss is interested in finding out about the probability of 10 acceptable cans in each "Zip Pak," Sims would specify that 10 of the 12 trials be "successes." As it is common knowledge 1 out of every 10 cans in the past has been defective, then there is a 10 percent likelihood the product will be defective in any given trial. Alternatively stated, there is a 90 percent probability of having an acceptable product in any given trial.

Armed with this information, Sims, demonstrating why he is paid the hefty seven-figure salary that he so richly deserves, would thumb through his *Statistics for Quality Control Managers* handbook (comic book version) until reaching the binomial probabilities table. He would run his finger down the first column (the number of trials) until reaching 12. To the right of the 12 would be a vertical row of numbers from 1 to 12 representing the number of successful trials or, in this case, the number of acceptable cans of Zippy Cola in the "Zip Pak." He would run his second finger down to the number 10 since the Great Boss is interested in the likelihood of pulling exactly 10 acceptable cans out of a given "Zip Pak." Sims would then move his third finger horizontally across a series of probabilities ranging from .01 to .99 and stop at the .90 column because of the fact there is a 1 in 10 chance in any given trial of getting a defective can of Zippy Cola. The number under the .90 column at that level is 230, which means, as Sims well knows, there is a .230 probability that exactly 10 cans in the 12-can "Zip Pak" will be acceptable.

What if the Great Boss is in a particularly demanding mood that day and wants to know about the lagging "Zip Hips," which features 3 five-gallon jugs of Zippy Cola attached to a belt, that can be worn around the waist? The Great Boss is stumped that sales of the "Zip Hips" (which weighs nearly 200 pounds when the three jugs are attached to the belt) are very poor. After all, he himself had developed a high-budget commercial featuring young models trying to drag their "Zip Hips" down city streets and up department store escalators. Not being very observant, the Great Boss wonders whether the 30 percent defect rate of the "Zip Hips" jugs is to blame for the public's indifference to the product. So the Great Boss calls Sims once again and asks him to determine the probability that all three of the jugs will be defective given the fact the latest quality control tests show any given jug will not work 20 percent of the time (or, that any given jug has an 80 percent probability of working). Sims retrieves his manual once again and soon determines, using his binomial distribution table, the probability is 0.008 that none of the three jugs in a "Zip Hips" will work at a given time. The probability that a wearer of the "Zip Hips" product will be able to walk comfortably any time soon is probably not much greater.

Although the binomial distribution is fairly clear-cut, the casual student of statistics might, at first blush, believe it would be fairly difficult to calculate the mean, variance, and standard deviation for a binomial experiment. Fortunately, this suspicion is unfounded. The mean for a binomial experiment is obtained by multiplying the number of trials by the probability of the event in any given experiment. If we flip a coin 1,000 times, for example, we know that the probability of the coin landing on "heads" each time will be 0.50. Thus we obtain the average number of predicted "heads" merely by multiplying the 1,000 trials by 0.50 to give us 500.

The mean, variance, and standard deviation are most helpful in the quality control area. Suppose the Goody Boy Cookie Company determines that only 70 percent of all of the "Big Chip" chocolate chip cookies actually have chocolate chips in them. This means there is a defect rate of 30 percent, which cannot be explained away as merely being "dietetic" cookies. In a sample of six cookies, the mean number of nondefective cookies would be equal to the number of cookies (6) multiplied by the probability of obtaining a cookie with chocolate chips (0.7), or 4.2. The variance here would be obtained by multiplying the mean (4.2) by the probability of defective cookies (0.3), or 1.26. Finally, the standard deviation would be found by taking the square root of the variance (1.26), which is equal to 1.12.

There are numerous ways in which a little knowledge about probability distributions can enhance one's life. Indeed, one can have a more pleasant vacation at a glamorous casino. Alternatively, one can seek to advance one's own career in any industry by showing his or her fellow employees how to calculate, among other things, the probability of selecting a certain number of working products from a batch that also contains a number of defective products. Those who have a more scientific bent can even set up their own experiments and test the effects of any defective product on whomever happens to be in the wrong place at the wrong time and thus help advance the frontiers of human knowledge.

Standard Normal Distributions

The number is certainly the cause. The apparent disorder
augments the grandeur.
—Edmund Burke

M ost books that talk about standard normal distributions do so in a very predictable manner. They present the concept of the standard normal distribution and then demonstrate the ways it can be used to solve a variety of problems. Suppose we wish to take a different approach, to entertain the reader and perhaps provide a bit of education. If we were interested in reaching a wider audience, such as the people who buy romance novels, then we might try to introduce the concept of the standard normal distribution in the following way:

Janet gazed into Rex's piercing blue eyes as he pressed his lips to hers. "Oh, Rex," Janet moaned, as Rex nibbled on her ear. "Tell me more about continuous probability distributions." Rex pulled Janet closer to him and she could feel his hot breath on her neck. Suddenly she was overwhelmed by his presence. "Oh, Rex," Janet gasped. "Your tongue is making me gag." Rex wiped his mouth with his sleeve. "Excuse me, dear Janet," Rex said. "I was swept away by your intoxicating beauty." She pulled him closer and felt the warmth of his chest passing through her fishnet stocking negligee. How she loved Rex. No one had ever made her feel so feminine and so alive! Except for Herbert. But Herbert had lost his mind and was now kept handcuffed in a sanitorium. Dear sweet Herbert, Janet thought sadly, as Rex placed her on the bed, his rippling forearm wrapped around the softness of her back. "Oh, Rex!" Janet begged. "Please give me your throbbing random sample! Give it all to me!" As Rex lowered his glistening naked body, Janet wondered if Herbert still was able to do hand puppet shows for the other patients. . . .

Certainly this more accessible approach would mark a bold new direction in the teaching of introductory statistical theory. It would probably be somewhat self-defeating because the lesson about the continuous probability distribution would tend to get lost in the lurid prose and erotic symbolism. The drawbacks associated with this sort of "statoerotic" approach give us little choice but to revert to the dry, dusty approach that has proven so fruitful for training generations of future statisticians.

Although we spent much of the previous chapter engrossed in a very erudite discussion about discrete probability distributions, we must move on to continuous probability distributions. This is not a move calculated to call greater attention to the study of statistics. Indeed, we shall spend much of the remainder of this book discussing various types of discrete probability distributions. This news may not cause many people to whoop with joy, but it will at least illustrate the ways fundamental foundations of statistical theory are grounded in these continuous distributions.

A continuous probability distribution is like a blanket in that it covers all possible values between the specified limits. We can think of the blanket's edges as the analogous numerical parameters that define the range of a continuous probability distribution. Perhaps the simplest type of continuous probability distribution is one having a uniform distribution. A uniform distribution is not very exciting to illustrate because its values do not change over the entire range of possible outcomes. One simple example of a uniform distribution would be a graph that showed the number of ticks emitted by a clock each minute on the vertical axis. The horizontal axis would show the passage of time in minutes. If the clock continuously ticked 60 times per minute, then we would draw a straight line from the "60" mark on the vertical axis and extend it parallel to the horizontal axis. The line would tell us that the clock was continuing to tick 60 times each minute for the entire amount of time shown on the horizontal axis.

Unlike a bell-shaped distribution function in which the values rise, peak, and then fall, the values of a uniform distribution remain the same over the range of possible outcomes. As such, a uniform distribution is graphed as a straight line, which is projected outward from the vertical axis. The area constituting a uniform distribution is a rectangle because it is bounded on both the upper and lower ends of its ranges by straight lines, which project upward from the horizontal axis. These lines are, in turn, connected by the uniform distribution. What makes the uniform distribution so exciting? Nothing, really! But it offers a very simple explanation of the way probability distributions are described. The uniform distribution can reveal the relationship between the area on the graph and the probability that a given variable will occur between two parameters.

Where might we find a uniform distribution? Suppose I was a teacher of very bright students at Muffy-Biff High School, and all ten students in my introductory economics course received perfect test scores on their final exam. I might want to create more of a Hobbesian atmosphere by telling my students that only the two students who gave me the most expensive teacher appreciation presents would receive the only As. Such an approach would not be fair, aside from being a bribe, because it would naturally favor those students who came from wealthier families. But I would remind the students that life itself is not fair, and that they should get used to uneven treatment in the workplace. However, I would probably want to give all the students As if only to forestall the inevitable lawsuit.

If I wanted to graph this grade distribution probability, the vertical axis of my graph would have a grading scale from 0 to 100, and the horizontal axis would be numbered in consecutive whole numbers to represent the number of students involved. Since every student received a perfect grade, the line would project outward from the 100 on the vertical axis mark and extend parallel to the horizontal axis until it reached the number corresponding with the total number of students in the class. Thus, this graph would show that every student in the class received a perfect grade on the test. If we shaded the area, we would find ourselves looking at another rectangle.

Unfortunately for those lackluster students who labor fruitlessly in the real world, our hypothetical class does not have much of a basis in reality. Grades in most classes are apportioned, based upon the performance of the students. There is usually a great discrepancy between the scholastic performance of the best students and the worst students. Our uniform curve would bring a smile to Karl Marx's face because it would show there was an underlying equality in the intellectual achievements of the entire class. Yet, the uniform curve would not be very realistic. No, our grade curve would reflect the differing interests and preoccupations of our diverse student body. Some of our students would be so serious that an A– would scar their psyches for life, whereas other, less academically inclined students would be pleased to carry a sturdy D+ average. We would expect our class curve to approximate, at least to some degree, the bell curve, with a few malcontents languishing toward the lower end of the qualitative scale (invoking their Fourth Amendment rights and bringing weapons to school) and the bulk of the students scoring somewhere in the average range, marked by the bulge of the curve. Finally, there would be the few students at the upper end of the curve who showed up at every class, did their homework assignments, dressed neatly, and did not slash the instructor's tires. It would be very unlikely to expect a uniform probability distribution for any given class of students due to the differing abilities and interests of the students themselves.

Standard Normal Distributions

The standard normal distribution is a function in which the values tend to rise, reach a peak, and decline to a lower level. These changes show there are relatively few values located at either end of a standard normal probability distribution; most of the values will cluster around the mean value. A graph of any standard normal distribution will be affected by two variables: the mean of the population and the standard deviation of the population. We can picture a horizontal axis numbered from 1 to 100 in increments of 10 that denote the mean value of the population we are studying. The mean itself would be found by drawing a line upward from the horizontal axis until it intersects at the highest point of the curve of the distribution. In any type of graphed standard probability distribution, the highest point of the curve will usually coincide with the mean value of the population. This makes sense when we remember that a standard normal distribution is symmetrical. In other words, the curve of the distribution is the same on both the right and left sides of the mean. A change in the population mean will cause the curve to shift to the right or the left along the horizontal axis. If the mean of the population declines, then the curve will shift left—back toward the origin. Similarly, an increase in the mean of the population will cause a rightward shift in the curve along the horizontal axis.

What about the standard deviation? Well, we should recall that the standard deviation is a measure of scattering or of the distribution of values in a population. The more the values are dispersed, the greater is the standard deviation. How does this affect the shape of our curve? Simply put, the greater the dispersion, the flatter the curve. If all the values in a population are clustered relatively close to the mean (peak) of the population in a normal distribution, then the curve will appear to be "pointed," rising quickly to the peak at a very steep rate of ascendancy and then declining just as quickly after passing through the peak. While this gold nugget of information should be enough to drive even the most statistically challenged reader into fits of ecstasy, we should point out this relationship works in reverse. When the dispersion of values in a population is greater, the curve itself is flatter and the rate of ascendancy of the curve is much less pronounced.

What is the standard normal distribution? It is the simplest and yet most elegant probability distribution. It has a standard deviation of 1 and a mean of 0. By all appearances, the standard normal distribution is symmetrical in shape—appearing to its wildly cheering fans as the perfect bell curve.

Mathematicians have spent many days and evenings calculating the probability corresponding to a particular area under the curve, which is bounded by any two numbers along the horizontal axis. For our purposes, we will not reproduce the tables of these values but instead provide the specific information to show how to calculate specific probability distributions. For example, the area under the curve bounded by values of 0 and 1 is 0.3413. This means about 34 percent (or .3413 if we wish to be exact) of all the values in the standard normal distribution lie between 0 and 1, which is also equal to one standard deviation from the mean. Similarly, the area under the curve bounded by values of 0 and 2 is 0.3413 plus 0.1359, or 0.4772. So we would expect to find about 47 percent (or .4772 if we wish to be exact) of all values in our distribution between 0 and 2, which is also equal to two standard deviations from the mean.

Because 0 is the mean of the standard normal distribution, it follows this curve must also have a left-handed side as well. We would soon find that the values for one standard deviation (0 to −1) or two standard deviations (0 to −2) are the same as those found on the right side of the mean (between 0 and 2). As a result, we would expect to find about 47 percent of all values in our distribution from 0 to −2. Using our flawless logic, we would quickly determine that about 95 percent of all values would lie within the range between −2 and 2 by adding the two probabilities (.4772 and .4772) together. When statisticians refer to those values lying two standard deviations from the mean, they mean that part of the curve bounded on one end by −2 and on the other end by 2.

Some statisticians are devil-may-care individuals who throw caution to the wind and ask about the number of values lying within three standard deviations of the mean (between −3 and 3). These hardy souls know approximately 99.7% of all values will lie within these parameters. Obviously, the shift from two to three standard deviations lacks a certain dramatic flair; the number of values encompassed within three standard deviations (as opposed to two standard deviations) increases by only about 4 percent. This means that nearly all of the area beneath the curve is contained within two standard deviations from the mean. From the graph's standpoint, the curve at this point has become nearly flat at both ends and is gently curving downward toward the horizontal axis over which it is graphed.

Some would wonder why we should spend this time talking about something so theoretical as a standard normal distribution when there are so many things in the world to be done? After all, what utility can there be in standing on a beach drawing bell curves in the sand? Aside from improving one's hand–eye coordination, the standard normal distribution can provide us with a great deal of assistance in determining the defects of many types of manufactured products.

Yes, Virginia, there is a practical application to the standard normal distribution. This usefulness becomes more apparent when we consider the difficulties inherent in manufacturing precise measuring instruments such as wooden rulers. Suppose the Splinter Wood Products Company has a stamping machine that imprints foot-long strips of cut wood with a 12-inch scale in increments as small as one-eighth of an inch. Unfortunately, this stamping machine is somewhat rickety and inaccurate in that the markings are not always precisely matched to the ends of the ruler. As a result, the markings sometimes begin at one-fourth of an inch or even one-half of an inch when they should properly begin at zero. This inaccuracy annoys Sid Zebbo, the owner of Splinter. Very few customers seem to have any interest in purchasing inaccurate rulers (despite Sid's insistence that the "shifting of the marks" is due to the tidal forces of the moon). Obviously, Sid would want to determine the probability that a customer would get a ruler with inaccurate markings.

If Sid determines that the normally distributed values have a mean of 0 (the endpoint of the ruler) and a standard deviation of one-eighth of an inch, then Sid may want to ask his statistical guru Larry "Three Card Monte" Spinebender to determine the probability that a customer could obtain a ruler with the markings shifted one-eighth of an inch. If the one-eighth-inch discrepancy is equal to one standard deviation, Spinebender would consult his trusty standard normal distribution chart and find that the value for one standard deviation (1.00) is equal to 0.3413. This means that about 34 percent of all the values in this probability distribution are contained within the area under the curve to the right of 0 and lying within one standard deviation of the mean. Because the area to the right of 0 is 0.5 (which represents one-half of the total area under the curve, the other half being to the left of the mean 0), we would subtract 0.3413 from 0.5, giving us an answer of 0.1587. We could then conclude there would be a 0.1587 (about 16 percent) probability of selecting a ruler whose markings were shifted in a particular direction by more than one-eighth of an inch.

What about those situations where the amount of the inaccuracy is not as tidy as one or more standard deviations? We live in a messy world replete with all sorts of errors and inconsistencies, so we must also know how to use the standard normal distribution in situations where the amount of the inaccuracy is some fraction of one or more standard deviations. To return to our stamping machine example, let us suppose that Sid Zebbo finally hires Otto Hertz, a one-time gardener turned stamping machine technician, to fix the mechanical problems. Sadly, Otto's method for fixing delicate machinery is pretty much the same as his method for removing weeds from gardens: He pulls out his hoe and hits the side of the machine like a madman until he is satisfied it is working properly. What if Otto's "tough

love" approach toward machine repair causes the cali-
brations to become even more skewed so that some of
the markings are now about one and one-half inches off
and the standard deviation is equal to one inch? How
would we determine the probability of pulling one of
these mismarked (by at least one and one-half inches)
wooden rulers out of the bin? Why, we return to our
standard normal distribution table and find that the
value for a standard normal distribution value of 1.500 is
equal to 0.4332. In turn, we would subtract 0.4332 from
0.5 (which represents that one-half of the standard
normal distribution curve to the right of the mean 0) to
get 0.0668. We would then conclude there would be a
0.0668 probability of selecting a ruler which is mismarked by more than one and one-half inches once the
stamping machine has been "repaired" by Otto.

What if Sid wants to take this analysis further (now
that Otto has displayed his unique mechanical skills) to
determine the probability of pulling out a wooden ruler
whose markings are displaced between one-half inch
and one and one-half inches when the standard deviation is equal to one inch? Bear in mind, we are essentially trying to measure the width of the slice of the bell
curve, which is defined by these two boundaries (one-
half inch and one and one-half inches). We already know
from our previous foray into deep thought that the
value of a standard normal distribution of 1.50 (one and
one-half inches where the standard deviation is equal to
one inch) is 0.4332. Yet, we are not seeking the probability of pulling a ruler with markings displaced by more
than one and one-half inches. Instead our interest is in
the one and one-half inches as the upper boundary of
the slice of the curve. We then go to our table and find
the value of the standard normal distribution of 0.5,
which is equal to 0.1915. To find the probability that the
inaccuracy in the markings of a wooden ruler is between
one-half and one and one-half inches, we subtract 0.1915

from 0.4332 and obtain 0.2417. The probability is 0.2417 that the markings on a randomly selected wooden ruler would be off by at least one-half but by no more than one and one-half inches.

Is the world such a perfect place that every normally distributed population has a mean value of zero and a standard deviation of one? No, not really. This fact forces us to consider another type of probability distribution function that is more suited for the rough-and-tumble real world—the nonstandard normal distribution.

Nonstandard Normal Distributions

Even though the nonstandard normal distribution may have a different (non-zero) mean value and a standard deviation less than or greater than one, we do not have to discard our standard normal distribution table. We are fortunate to have an equation that will allow us to standardize the value of these nonstandard distributions:

$$z = \frac{x - \mu}{\sigma}$$

In this equation, z represents the number of standard deviations that a value x is from the mean value of the population. The value z is known as the standard score. The symbol σ represents the standard deviation of the population. This equation thus permits us to wade into the quagmire of nonstandard distributions without a care in the world, knowing full well these problems will yield to our relentless investigations.

As with the previous equation for standard normal distributions, this equation can be solved once we have determined (1) the value x; (2) the mean value of the population; and (3) the standard deviation. Because this equation can be used in nonstandard normal distributions, it has a much wider applicability (and, hence, utility) than the equation for purely standard normal distributions.

Suppose we have been sent by the King of Wittenburg, a small principality in Central Europe, to find a competent hair stylist to care for His Majesty's golden locks. Our search leads us to the doors of the Ludwig von Beethoven Academy of Hair Management, the world's foremost school of hair design and coloring. As we have little time to waste, we meet with the headmaster, Mr. Beezel, and ask for some names of his top students so that we may interview for the position of Royal Hair Stylist. Mr. Beezel pulls out a list of the previous semester's grades and determines the mean score was 100 and the standard deviation was 20. If we want to determine the probability of selecting a student whose grades lay somewhere between 100 and 160 (three standard deviations), we would be looking for the shaded area under our bell-shaped curve under which was bounded on the horizontal axis by the numbers 100 (on the left) and 160 (on the right). By subtracting the mean score (100) from the upper limit of our parameter (160) and dividing the total (60) by the standard deviation (20), we will arrive at a z score of 3. When $z = 3$, our standard normal distribution table tells us the shaded region of the curve is 0.4987. As a result, we have a nearly 50 percent probability of selecting a score between 100 and 160. Presumably, His Majesty will want a slightly more exacting method for finding the new Royal Hair Designer.

How much could we pare down our applicant pool by only focusing on those students whose scores fell between 140 and 160? To find the answer to this question, we must return to our standard normal distribution table and find the area under the curve bounded on the left by 140 and on the right by 160. Our table tells us the value for 3 standard deviations is 0.4987 and the value for 2 standard deviations is 0.4772. We calculate the area under the curve between these two boundaries by subtracting 0.4772 from 0.4987, which gives us 0.0215. At that point, we will pat ourselves on the back because we know we will be dealing with the top percent of the students at the Academy. It is very unlikely any one of these students will scald His Highness's scalp with chemicals or turn his golden locks a motley orange.

How do we determine the probability of randomly selecting a score when the standard deviation is not an integer (a whole number such as 1, 2, 3, etc.)? Fortunately, we are still able to use the same equation. To return to our hair stylist example, we might find our efforts to land a top-notch student to work in His Majesty's court would be more difficult than we originally anticipated because the best students had already been snatched up by various Arabian sheiks and chairs of multinational corporations. We could be left with the problem of trying to determine the probability of selecting an above-average as opposed to a superior hair stylist. We might find the scores of the remaining students in the upper half of the graduating class range from 100 (the mean) to 127 (the highest score). If the standard deviation is 20, we subtract 100 from 127 (which equals 27) and divide 27 by 20, giving us a standard score of 1.35. If we look at our standard normal distribution table, we find the standard score is 0.4115. So we have a probability of 0.4115 of selecting a student who has a score between the mean value of 100 and the highest value of 127 in the fair-to-middling student population.

If we decide to focus more narrowly on the upper end of the remaining students, we might want to determine the probability of selecting a student whose score falls between 120 and 127. We already have the value of 0.4115, which gives us the area under the curve between 100 (the mean score) and 127. We want to subtract from that area the area lying between 100 and 120. We know this area is equal to 20, which is also equal to 1 standard deviation. Our table tells us the area under the curve equal to 1 standard deviation is 0.3413. We then merely have to subtract 0.3413 from 0.4115 to get the area under the curve between 120 and 127, which is equal to 0.0702. This excursion in mathematical mayhem tells us that approximately seven percent of all the students scored between 120 and 127 at the Academy. This may or may not be a sufficiently large enough pool of applicants for us to interview for the court position.

The usefulness of nonstandard normal distributions also becomes apparent when manufacturers must consider the needs of their intended buyers. Suppose the Boll Weevil Clothing Company decides it wants to market a new line of clothes designed by the noted colorblind fashion magnate, John des Fleurs, for those free-spirited individuals over six feet tall who like to combine such things as red or green plaid shirts with yellow or blue striped or polka-dotted trousers or skirts. Leaving aside the question of whether a designer who proposes such a clothing line should be shot by a firing squad for "crimes against humanity," the marketing director of Boll Weevil needs to determine whether it makes sense to essentially ignore those potential customers under six feet tall. This concern would be prompted not so much by fear that some of the sub-six-footers will take offense by not being able to purchase such exciting garments. No, the issue is one of profits and losses. The marketing director does not want to plunge ahead with a very expensive advertising campaign that cannot be justified given the size of the intended audience.

With this issue set squarely in place, the marketing director may want to call in the company statistician to determine the size of the six-footer audience. The company statistician may in turn thumb through any number of statistical abstracts to get some idea of the six-footer audience. Before he gets much further, the powers-that-be at Boll Weevil may decide they are not interested in making John des Fleurs originals available to people over seven feet tall. This limitation might be due to the perception there are too few people over seven feet tall to justify the expense for manufacturing clothing for them. This perception might also be prompted by the marketing director's pathological hatred of seven-foot people (due to an incident at a basketball game where a seven-foot player took exception to the marketing director's opinions regarding the player's position in the food chain and threw a basketball in his face). In any event, the marketing director will want to know how much of the potential market for the John des Fleurs line will be lost by targeting only those customers between six and seven feet tall.

This brings up a problem we have dealt with before because it essentially requires us to determine the area under a bell-shaped curve that is bounded on one side by individuals six feet tall and on the other by individuals seven feet tall. The statistician will want to look at the population as a whole and determine the average (mean) height of the population and the standard deviation.

What if the statistician discovers that the mean height of the population is 68 inches tall with a standard deviation of three inches? This means he would have to start more than one standard deviation above the population mean before he came close to the beginning of his targeted audience. Based upon our past experience with normal distributions, we know this is not a good sign if we are hoping to target our products at the widest possible audience because our table tells us that one standard deviation is equal to 0.3413. As a result, we know we have already lost about 85 percent of the population as a whole as we are necessarily excluding the 50 percent of the population whose height is below that of the mean value. If the six-foot mark lies 1.33 standard deviations above the mean, our table tells us that only about ten percent of the population will be within our targeted market. Our table informs us 1.33 standard deviations is equal to 0.4082. As we are dealing with only one-half of the normal distribution curve, we subtract 0.4082 from 0.5 based upon our standard normal distribution table to get 0.0918 (our proverbial ten percent). Naturally, this will cause us some concern because we are targeting our advertising at only ten percent of the entire population.

What about the exclusion of those seven-foot tall consumers? Could we bolster our market share by manufacturing clothing that could be worn by people as tall as eight feet? Probably not. Based upon our prior information, a seven-foot tall person would be more than five standard deviations greater in height than a person of average height. This poses a bit of a problem for our planned television campaign to welcome people who cannot walk through doorways without bumping their heads. There are simply not very many seven-foot-tall individuals. The fact they might look very sophisticated in original John des Fleurs garments would not disguise the fact all the seven-footers in the country would add only a minuscule number of consumers to the targeted audience. After all, we would be excluding more than 99 percent of the audience if we moved merely three standard deviations above the mean of the population. As a result, we would not find ourselves deluged by legions of seven-footers at fine department stores throughout the country. It would be too difficult to form a single legion even if we could get them all together in one place.

The Central Limit Theorem

In continuing our quest for knowledge and understanding, we now turn our attention to one of the most significant tenets of statistics—the central limit theorem.

It holds that if we go out and collect numerous samples of the same size, determine their mean values, and graph them, then they will tend to assume a bell-shaped curve that approaches a normal distribution. This sort of revelation will not only allow us to sleep more soundly at night but also illustrates that statistical samples—like populations as a whole—will tend to more closely approximate the bell curve shape of a normal distribution as the size of the sample increases.

Although Hamlet will never be heard praising the central limit theorem, the theorem does offer a certain profound insight into statistics. If we took the test scores of 100 students in the hair-fluffing class at the Beethoven Academy of Hair Management, we would not be surprised to see they tended to approximate a normal distribution with comparatively few scores in the lowest ranges (those who could not fluff hair even if they were using an electrical wire with 10,000 volts) and the highest ranges (those who could fluff hair through pure thought alone). Indeed, we would find that the bulk of the scores were concentrated in the middle area (ranging from barely tolerable fluffers to semi-professional fluffers).

This discovery that most of the class is fair-to-middling would not surprise us because that is the fundamental conclusion to be drawn from any type of normal distribution in which a quantitative evaluation (such as grades or test scores) is used. The central limit theorem says we can take those 100 scores, randomly allocate them among ten different samples, find the means of each of those ten samples, and still construct something approximating a normal distribution with those mean scores. This conclusion is not at all obvious because we tend to believe that the normal distribution begins to resemble a bell curve only after we are wading knee-deep in measurements. The idea that we can keep shuffling these 100 scores, tossing them randomly into ten different piles in any way we choose, and then construct what appears to be a normal distribution of sample means is quite remarkable.

Here we are merely concerned with understanding the idea behind the central limit theorem. Statisticians have found the central limit theorem does have a few wrinkles in that we usually need to have a sample size of at least 30 before the sample can generally be expected to approximate a normal distribution. Not surprisingly, this emulation of the bell curve will become even more pronounced as the sample size increases. We can save some time if we are dealing with a normally distributed population because the sample means will always be normally distributed regardless of the size of the sample sizes. We will also find that the larger the size of the samples, the less variation there is in the sample means. This conclusion makes sense if we think about our 100 student scores. We are much more likely to have sample means that are at one end or the other of the normal distribution curve with 10 sample means having 10 scores each than we would if we had 5 sample means with 20 scores each. The high-fliers or the bottom-dwellers will tend to get swept up into a larger group of fair-to-middling scores so that the dispersion of the means decreases.

The wonder a young child feels about the world is obviously tapped when an adult learns about continuous probability distribution functions and the standard deviations of populations. Of course, there may be a few things in life that will pack a bigger thrill than learning about both standard and nonstandard normal distributions. Yet this chapter, like its predecessors, should help the reader learn a little more about the ways statistics makes it possible for us to impose order upon the apparent chaos of our universe.

Sampling the Samples

Many small make a great.
—GEOFFREY CHAUCER

W e could not have failed to notice the great reliance that even the most brilliant statisticians place on accurate sampling techniques. After all, we all use sampling techniques of one sort or another in our daily lives to help guide us in our decisions. There are very few of us, for example, who have not furtively snuck a few grapes in the produce aisle of the grocery store to assess whether the purchase of an entire bunch would be justified. This is a simple example of how people sample an item in order to learn about the entire population. In this case, we are taking a few grapes when the stockboy's back is turned so we can see if the grapes are tasty enough to merit their purchase. Of course, we may enjoy our sampling so much we get filled up and no longer have any desire to purchase grapes. We still would have obtained important information about such things as the tartness or sweetness of the grapes, the extent to which they are covered with ants, the aroma of the pesticides with which they were sprayed, and the roughness of their handling in shipment, based upon the amount of bruised fruit.

We also use sampling techniques when we engage in the search for that perfect mate. When we begin our search, armed only with the images of taut bodies provided by leading fashion and fitness magazines, we may find that the population of possible mates does not exactly correspond in appearance with centerfolds and cover models. Our first conclusion, as we search various places frequented by members of the opposite sex, is that the samples in the magazines we receive in plain brown wrappers are not representative of the general population. As a result, we might, in the finest tradition of statisticians everywhere, conclude that our sample is not representative of the population as a whole.

Having lowered our sights a bit and perhaps accepted the idea that muscular and shapely people are shallow and that we truly want to meet someone who is plump or lumpy, we might then begin to search for that certain perfect person. So we would begin our sampling efforts in earnest, unbowed by any concerns that we may never find that perfect person. How might we conduct our sampling process? First, we would have to decide where we need to be to meet people of the opposite sex. If we are trying to meet ladies, we might want to go to popular restaurants or nightclubs or even (if we thought we might be interested in an intelligent conversation) bookstores. Once we spotted one or more eligible women, we would then need to learn a little about them to assess whether they might be suitable mates. We could try to engage in pleasant conversation and discuss (with as straight a face as possible) such things as getting in touch with our feminine selves and being more sensitive. A more direct approach would be to walk up to each of the ladies and kiss them passionately on the lips. This is the type of unexpected maneuver designed to provoke a variety of responses right away and possibly lead to misdemeanor charges. The advantage of such a daring move, however, is that it forces the issue of whether the woman has any interest in seeing you again to the forefront. As she has never seen you before, she may react with some hostility and try to hit you. Or she may be so taken with your romantic parry and thrust that she will pretend, with at least some sincerity, to be interested in you. Either way, you will soon be on your way to finding that perfect person with whom you can spend your entire life—or at least a few years until your midlife crisis.

Although this example offers some interesting similarities to statistical sampling techniques, it would not be a true sampling technique unless we were using our knowledge about the various women we talk to (and, on occasion, kiss) to make inferences (or draw conclusions) about the population of women as a whole. If we were simply trying to date women and not learn anything about them as a group, then we would not really be engaged in a sampling process to which most statisticians could relate.

The utility of inferential statistics is totally dependent upon the quality of the known sample data because we will necessarily use that sample to make inferences about characteristics of the population we are interested in studying. Suppose we are in a particularly inquisitive mood and wish to determine the mean value of a population, such as the average age of a doughnut in a Manhattan coffee shop. Certainly we could tackle this job the hard way by hailing a cab at Battery Park and having the cab take us to each and every coffee shop in New York City. We could then personally determine the age of each doughnut and, after adding the various ages together and then dividing by the number of doughnuts, calculate an average age. Of course this plan presupposes we have a reliable "Do-Nut-Meter" to give us accurate readings as to the exact ages of each of the sometimes tender, sometimes rock-hard treats. Even if we did have such a device, however, the question remains as to why we would want to spend all that time (never mind the cab fares) rushing all over Manhattan to do such a ridiculous thing. Well, there are some health-related issues that do come into play. After all, we do not want to allow coffee shop proprietors to sell month-old doughnuts to the public even though they may be a delightful shade of bright green and suitable for gnawing. It still seems like a lot of work to send a detail of professionals all around the city in search of doughnuts, particularly when a doughnut's perishable nature ensures the search will be a never-ending task.

If we do not wish to outfit several doughnut patrols to be on 24-hour duty throughout the metropolis, we must come up with a more efficient and effective solution. We might seek to bring in the heavy hand of government by banning the production, distribution, and consumption of doughnuts altogether but this is, at best, a limited solution because there will always be other kinds of foods we will want to monitor for freshness. Moreover, there are many people who would not hesitate to take part in an armed insurrection if we tried to enforce such a drastic measure and ban nature's most perfect food. In the interest of avoiding civil war and becoming a target for sugar-deficient snipers, we could take the more intelligent approach of sending our "doughnut scouts" out to a certain minimum number of coffee shops to estimate the average age of the doughnuts in this sample. How many coffee shops should we travel to as part of our investigation? One? Two? Ten? Twenty?

How we decide the number of doughnuts to include in our sample will depend in part on the size of the population as a whole (the number of doughnuts in Manhattan) and the confidence we have in the representativeness of the sample itself. We might, on our first venture into the city, stumble upon the one doughnut shop with doughnuts having an average age of the doughnut population as a whole. However, that happy result is very unlikely and we certainly will have no way of knowing the accuracy of our sample when we sit down for a cup of coffee at the counter. Prudence dictates we expand our sample size so we have a better chance of approximating the average age of all the doughnuts in Manhattan.

If there are, say, 2,000 coffee shops in Manhattan, each with an average of 500 doughnuts, then it is fair to say our population would be equal to about one million doughnuts. If we assume, for purposes of this example, that doughnuts can be acquired only at coffee shops (as opposed to grocery stores or pushcarts), then it simplifies our investigation of doughnut maturities. Given the fact that one shop has 500 doughnuts, we might be tempted to conclude we need only go to the one place and calculate the ages of those very same 500 doughnuts. After all, that would give us a sample size of one-twentieth of one percent, which is not so very small when we compare it to the electronic sampling used to identify the most popular television shows in America. We might want to be careful because of the fact we would be relying on the stock of a single doughnut shop. There is some danger we might choose Ye Olde Doughnut Shop, the coffee shop with the oldest doughnuts in Manhattan, a fact that is a source of pride to its owner, a food preservation enthusiast. This is a situation that will cause our calculation of the average age of the doughnut population to be wildly exaggerated. Or we might stumble across Fred's Fleeting Pastry Shop, where pastries are thrown away if they remain unsold for more than ten minutes due to the owner's paranoid belief that huge furry germs will nest in any doughnut not consumed within a few minutes after being pulled from the fryer. Certainly the doughnuts in Fred's store, to the extent we would even be able to get a hold on them, would have an unusually tender average age that would be far less than the Manhattan doughnut population as a whole.

Complicating our situation would be the fact that the funding available for the Do-Nut Meter might be only enough to allow for testing 500 doughnuts. This testing constraint would obviously prevent us from testing tens of thousands of doughnuts. However, we should probably regard this limitation as a blessing in disguise because we would not want to be tempted to begin racing through Manhattan with big empty sacks to collect such vast quantities of doughnuts. The obvious need appears to be to avoid relying on a single coffee shop while still limiting our sample size to the 500 doughnuts, which we know we can test using the Do-Nut Meter. The equally obvious solution would be to decide upon a representative number of doughnut shops to visit and, in turn, a representative number of doughnuts to sample at each of these shops. We might, for example, decide to go to 50 coffee shops throughout New York City and test 10 doughnuts at each shop, or we might decide to go to 20 coffee shops and test 25 doughnuts at each place. There may be no single best solution here for determining the composition of the sample; we merely want to get a representative sample of doughnuts from a representative sample of coffee shops. Regardless of whether we choose 10, 20, or 50 coffee shops, we will ultimately be faced with determining the average age of the doughnut population.

Assuming that we finally gather together 500 measurements of doughnut ages, we will then have to choose which statistical measure to use to estimate the mean age of all the doughnuts in Manhattan. We could find the median or the mode value of the sample. Each of these values would provide us with a fairly accurate estimate of the mean value of the population. Fortu-nately, the danger of making a wrong decision here as to whether to select the sample median (the doughnut age halfway between the youngest and oldest doughnuts), the sample mode (the doughnut age that occurs most frequently in the sample), or the sample mean (the total age of all the doughnuts in the sample divided by the total number of doughnuts) is quite small.

Certainly we would not find ourselves in mortal danger if we were to select the wrong measure. Nor would an incorrect decision result in a great financial loss unless we were given that question while playing in the final round of America's least-watched gameshow, Statistical Challenge (thus costing us a lifetime supply of pocket protectors). We would probably not be very surprised to learn that the sample mean is, in most cases, the best estimate of the population mean. Of course the size of the sample and the values contained within the sample will affect the accuracy of the sample mean as a predictor of the population's mean. On the whole, the sample mean is a fairly reliable estimate of the population mean because it tends to have less variance around the population mean than does either the sample median or mode values. We should not be surprised to find that the sample mean provides us with the next best thing to the population mean. The sample mean is certainly much more convenient. We can get the sample mean of our doughnut population by venturing into 20 or so coffee shops scattered throughout New York City as opposed to sending highly trained "pastry police" into every single coffee shop on a precisely coordinated mission to inventory the ages of every doughnut in Manhattan.

Suppose our group of plucky scientists selects 500 doughnuts ranging in age from 5 minutes to 27 weeks and determines that the mean age of the 500-doughnut sample is 4.5 days. This conclusion might be viewed by the participants as the grand culmination of their life's work or it might be regarded as an indication that they should instead have a bagel for breakfast. The mere fact that the scientists had estimated that the average age of all the doughnuts in Manhattan is 4.5 days would not be the final issue to be considered. After all, how could they be sure this number had any real validity? Alternatively stated, would it be prudent for these scientists to have a high degree of confidence in the accuracy of the sample mean measurement of doughnut age? As with most things, the simple answer is "maybe."

In statistics, the level of confidence has nothing to do with the swagger in a statistician's walk, or the amount of weight he can lift over his head, or even the levels of testosterone in his blood. Instead it is a concept that relates to the quality or accuracy of a statistical measure, such as our sample mean of the Manhattan doughnut. How can our researchers be confident that this sample mean is truly representative of the doughnut population as a whole? Quite simply, our researchers can retrieve their elementary statistics textbooks and turn to the chapter on sampling, where they can read amidst the tales of murder, lust, and greed about such things as margins of error and degrees of confidence.

The margin of error has nothing to do with the programming choices of network television executives or even the effectiveness of the "rhythm method." Instead it has to do with the probability that the value of the sample mean differs from the value of the population mean. While it would be very entertaining to review in detail the formula for determining the maximum error of the estimate for a population mean, we shall have to be content with the knowledge that the margin of error provides us with an evaluation of the accuracy of our sample.

A related topic is the degree of confidence, which is the probability or percentage value that the parameter of the population mean is expressed in a given confidence interval. When we see television commentators droning on and on about degrees of confidence, they are usually referring to a measure of reliability that the confidence interval contains the population mean. Accordingly, they may say that the degree of confidence is 90 percent or even 99 percent. Whether we choose a value closer to 90 percent or 99 percent will depend on how we wish to trade off between the precision and the reliability of our calculations.

As our prose is probably not prompting many people to jump up and shout "Eureka" we should step back a second and consider what we mean when we are discussing the margin of error. For the sake of convenience, we shall use "E" to refer to the margin of error. We also need to dust off our bell curve to explain how this concept works when we are considering the standard normal distribution represented by any such curve. We can imagine a bell curve divided in half by a vertical line that passes directly through the highest point on the curve, as shown in Figure 1. The horizontal axis or x-value of this line is the population mean, which is labeled with the Greek letter "μ". If we move both toward the left and the right along the horizontal axis, we will cover a certain distance E, which is the margin of error. If we venture further to the extreme right or left end of our bell curve, passing beyond the horizontal markings along the x-axis, which designate the value for E both to the left and the right of the population mean μ, we are essentially leaving the middle of the curve and venturing into the tails. The middle area of the bell curve (excluding the tails) is described by statisticians as being equivalent to $1 - \alpha$. Another Greek symbol? Is this discussion leading us into a thinly veiled classroom lecture on statistics? No, not really. We have no plans to put on a tweed jacket, light a pipe, and fumble around for our yellowed notes and amuse anyone with anecdotes

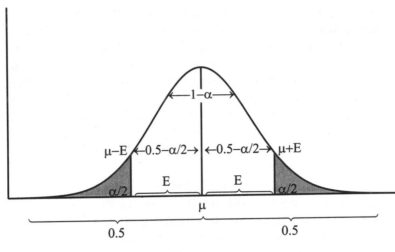

Figure 4.

about famous statisticians who moonlighted as professional wrestlers. Before we get too bogged down in this morass, we need to slap our faces with a bracing handful of cold water and remind ourselves that the bell curve itself is merely a graphical representation of probabilities. The number 1 is the sum total of all the values under a given probability distribution and means that the event is located within the boundaries of that curve. If the probability is less than 1, then it means we are not dealing with the entire area under the curve.

If we move along the horizontal axis of the bell curve, either to the left or to the right of the population mean μ, until we have covered the distance E in either direction, then we can describe the area under the curve bounded by $\mu - E$ and $\mu + E$ as the confidence interval limits. No doubt this is not particularly intriguing to you unless you like to do nothing more than curl up by the fireplace with a loved one and browse through a rousing book on statistical theory. But this area enables us to describe that part of the curve that is within our maximum estimate of error.

This may be a good time to think about the concept behind the bell curve. We need to recall that it provides a graphical description of a population and shows how that population is clustered based upon certain variables. Perhaps this would be easier to explain using the backdrop of a romantic evening. Suppose Rex is visiting Janet and has brought her candy and flowers. As this is not Janet's birthday or Christmas, we can only presume that Rex is in a very romantic mood. Now let us use this lovely setting to illustrate the concept behind the bell curve. If we imagine a two-dimensional graph in which the horizontal axis is represented by time and the vertical axis is represented by Rex's pulse, then we can depict the general course of the evening by charting the changes in Rex's heartbeat.

In the beginning, Rex's heartbeat is normal. As we move along the horizontal axis with the passage of time, his heartbeat remains at a fairly low level. As Rex and Janet embrace and kiss passionately, his heartbeat increases. On the bell curve, we would see the line describing his heartbeat moving further above the horizontal axis. As Rex and Janet became totally intertwined in each other's arms, the intensity of Rex's heartbeat would reach a maximum peak. When Janet checked the clock and saw that the shows for the new television season were beginning (thus checking Rex's final amorous moves), Rex's heartbeat would begin to decline, finally approaching the normal level, after a few minutes of catching his breath and massaging his disappointment. The curve describing his heartbeat over time would resemble a bell curve and would illustrate the way certain events cluster around a center. Here the curve would not be so pronounced because we would be dealing with the rate at which his heart beats. Unless Rex suffered a heart attack, this curve would be fairly flat. Even at his most intense moments with Janet (which Rex promised would be the "best two minutes" of her life), Rex's heartbeat would not increase by a factor of two or three.

But what lies outside the area $\mu - E$ and $\mu + E$? Quite simply, it is the extreme left and right ends of the bell curve, known by schoolchildren around the world as the "tails" of the bell curve. We can picture the "tails" as being the areas under the curve located at each end of the bell curve. The Greek letter α comes into our discussion because statisticians use it to represent the total area of the two tails under the curve. Because each tail covers the same area of curve, for purposes of this discussion, the area of each tail may be obtained by dividing α in half. This means that the value for the tail on the left end of the curve is given as $\alpha/2$. Similarly, the value for the tail at the right end of the curve is equal to $\alpha/2$. If we add the area of the two tails together, we obtain the value of α. Because α represents that area of the bell curve lying outside the maximum estimate of the error of the population mean E, the area within the two tails may be described as $1 - \alpha$.

Now that we have attached all sorts of pompous symbols to our bell curve to make ourselves feel far superior to those who spend their time reading tabloids, we can proceed to the next step and actually attempt to apply the concepts embedded in this curve. Remember we are attempting to determine whether the population mean lies within certain boundaries already defined as that area of the curve between $\mu - E$ and $\mu + E$. Since we are not including the entire bell curve, we cannot say at the outset that we are truly certain the value of μ lies within these boundaries. Instead we are focusing on the confidence interval to give us a probability that the value for μ is between $\mu - E$ and $\mu + E$. While this procedure may not make for good television, it is extremely important to statisticians and to industry as a whole. We need to be clear on what we mean by a confidence interval. We can be absolutely sure that our confidence interval limits will contain the population value μ only if they span the width of our normally distributed population. Because this does not really add a great deal of insight to our analysis we use confidence interval estimates to provide greater precision to our estimate of the mean value of the population. In short, we will be able to specify the degree of confidence (such as 90, 95, or 99 percent) that the population mean is contained within the boundaries μ + E and μ − E.

How can you learn more? Well, you could spend thousands of dollars and attend an exclusive university for four years and receive a degree that will lead you into the exciting world of fast food crew management or (dare we say it?) perhaps help you land that coveted sales associate position at the local department store. You can get a much greater bang for the buck by plowing through the remainder of this chapter and learning more than anyone else in your neighborhood about confidence intervals.

Suppose we want to know the average number of plaid burgundy pants owned by the members of the local rocketry club. We could use the tried and true approach and actually ask every one of the club's 78 members to reveal the number of plaid pants they own. We might find such a task too time consuming and the sheer drudgery of pretending to be interested in the most recent theories of rocket propulsion (as expounded at length by each of the club's members) to be unbearable. As a result we might want to select ten of the club's members at random and ask each of them to specify the number of plaid pants they own.

Suppose the average value of our sample is 7.8. This number is our point estimate of the mean number of plaid pants owned by all the members of rocketry club. We can already see we have no real indication that this estimate of the average number of plaid pants owned by every member of the rocketry club is very good at all.

Indeed, it could be a very rotten estimate. We might have happened to come across the only members of the rocketry club who actually wear pants and discover, much to our chagrin, that the rest of the club members are all Scotsmen who wear kilts. Thus we would get a greatly overstated mean value. Or we might find that the members of our sample group are chronic under-achievers in terms of their ownership of plaid pants and that the actual mean for the club members as a whole is much higher. We might have a few club members who believe they are the modern incarnation of the story-book emperor who bought new clothes and thus feel duty-bound to purchase a different pair of plaid pants for every day of the year. Certainly either one of these situations could prove that our sample size was too small to provide us with a good approximation of the population's mean value.

We shall bring together all the disparate (some would say incoherent) thoughts that we have consid-ered thus far in this discussion and actually construct a statistical model to wow the critics. To return to our plaid pants problem, we know that our sample size is 10 (10 members) and that the average value of the sample mean is 7.8. If, for the sake of argument, we assume that the standard deviation of the sample is 7, then we can find both the maximum error of the population mean estimate and the confidence interval for that very same mean value. We recall that α represents the area at either end of the probability distribution so that $1 - \alpha$ gives us the area underneath the curve bounded by $\mu + E$ and $\mu - E$. If α is equal to 0.05, then the z value (obtained by referring to the standard normal z distribution table in most statistical books) of $\alpha/2$ is equal to 1.96. The maximum error of the estimate of the population mean can be found by multiplying 1.96 by the sample stan-dard deviation (7) and then dividing that amount by the square root of the number of persons in our sample group (10). This calculation gives the resulting value for E of 4.34. If the average value of the sample size is equal to 7.8 and the value of E is equal to 4.34, then we can determine the confidence interval for the population mean. Do not allow the joy you feel to cause you to lose your focus and thus fail to complete the final stage of this operation. We must wipe our tears away and press forward, confident in the knowledge that we will soon know how accurate our sample mean is relative to the population. We see that the confidence intervals are defined by 7.8 + 4.34 and 7.8 − 4.34. This, in turn, appears to tell us that the value of the population mean will, in most cases, be somewhere between 3.46 and 12.14. However, this is not the right way to express this thought because we know that the mean value of the population (average number of plaid pants owned by the members) does not change. We are measuring the total number of pants and dividing them by the total number of members at a single point in time so that our data will not be affected by the subsequent efforts of a few disgruntled members to muddy our calculations by selling all their plaid pants to stores catering to fashion-challenged customers.

We need to look at this parameter provided by the confidence intervals in a different way. We are selecting data with a 95 percent degree of confidence. These confidence intervals tell us that if we continually select random samples of ten members and query them about the number of plaid pants they own, we will find that over the long term, 95 percent of the intervals obtained with each of the samples will in fact contain the value of the population mean. We should also keep in mind that the confidence interval is basically an expression of probabilities. Different samples of members will produce different plaid pants values, which will, in turn, produce different confidence intervals. We can expect that, over the long term, about 19 out of every 20 of the confidence intervals we derive will contain the mean value of the population. Because the mean value of the population does not change, each confidence interval will either contain or not contain the mean value of the population. Unfortunately, we will not know whether a particular confidence interval contains the mean value of the population unless we actually determine that number by obtaining the appropriate information from every single member. This sort of uncertainty is absolutely maddening to those who like to have everything in its proper place and who go so far as to stack neatly cornflakes in their cereal bowls each morning. The crux of the problem is that we simply cannot know with any certainty whether a particular confidence interval actually contains the population mean. We need to recall that the entire science of statistics is concerned with probabilities—not exact figures. As such, we are interested in finding out the ways things are most likely to be most of the time, so exact numbers are not necessarily our most important objective.

Now that we are feeling somewhat more confident about confidence intervals and margins of error, we can take another stab at using these concepts. Suppose we are interested in estimating the average number of mistresses or paramours of each member of the United States Congress. The somewhat sensitive nature of this research project would probably preclude our being able to obtain any government funding. Still, we might choose to proceed anyway to satisfy our own intellectual curiosity and possibly take advantage of the enormous opportunities for blackmail.

We would then set out to gather our information in any way we could. We are high-powered statisticians, so we would not be interested in querying every member of Congress. We know we should be able to obtain a fairly accurate estimate of the mean value of the population as a whole by selecting a representative sample group. We know there are 435 members of Congress. We might want to focus on a select 30 members of Congress for our sample, because 30 represents almost 7 percent of the total population of Congress. Plus, we know this research will not exactly endear us to the research subjects, especially after we begin publishing "scholarly," but revealing, photographs in major magazines.

Once we have randomly selected 30 members of Congress, we would have to determine the average number of lovers each member can currently claim. One particularly useful approach is that of direct face-to-face questions, usually accompanied by a video camera stuck in the face of the particular congressman. This approach can prove dangerous for the questioner. Furthermore, it does not usually encourage the respondent to answer with the utmost truthfulness, particularly if he or she is planning to run for reelection on a morality platform. Assuming that none of the 30 members of Congress will agree to speak with us about the number of lovers they are currently seeing, then we will have to resort to more clever approaches that do not rely on the forthrightness of our subjects.

We might want to order each of our researchers to follow a specific member of Congress and keep an ongoing record of his or her whereabouts. If we assume our researchers are a plucky lot who will do anything, including peering through apartment windows at night with night-vision goggles, wiretapping telephones, and intercepting heavily perfumed notes to ascertain the identities of congressional bedmates, then we can rest assured it will only be a matter of time before we have a fairly reliable database.

Let us assume, after several weeks of relentless investigation, we are able to determine our 30 legislators have a total of 73 lovers. We further find that some are monastic in their tastes whereas others prefer more exotic pleasures. In any event, we now have an opportunity to estimate the average number of lovers of each of the members of Congress.

As we are interested in striking a balance between precision and reliability, we would probably choose to proceed using the 95 percent degree of confidence we previously used. Given the fact each legislator in our sample has an average of 2.43 lovers (73 lovers divided by 30 legislators) with, say, a standard deviation of 3.3, we would multiply 1.96 (the standard z value corresponding to an area of 0.4750) by the sample deviation of 3.3 and then divide that total by the square root of 30 (5.47) to give a maximum error of estimate E of 1.18. We would then be able to determine the confidence interval by adding and subtracting 1.18 from our sample mean of 2.43. This would give us a confidence interval ranging from 1.25 to 3.61.

As with our burgundy plaid pants example, this information would provide us with the following insight: If we were to continue our investigations by collecting additional sample groups of 30 legislators, we would find over time that 95 percent of the corresponding intervals would contain the average number of lovers of all the members of Congress. We would not be able to say for sure that our sample average was the population average without getting the desired information from every member of Congress. We would know that 95 percent of all of those confidence intervals would contain the actual mean of the population.

Finding Your Sample Size

Finding confidence intervals is one way individuals can sharpen their mathematical skills and be the life of any party. Sometimes we need to determine the size of the sample needed in order to determine results with a given degree of confidence. We may not want to arbitrarily select a sample size as we did with our plaid pants and Congress examples but, instead, determine the most appropriate sample size given the constraints of our experiment.

How can the ability to predetermine the size of the sample be beneficial to statistical experiments? One obvious benefit is that we may have a limited research budget due to some lingering resentment over our previous congressional project. Indeed, a lack of cash could force us to make all kinds of revisions in our research methods. If funding cutbacks necessitated our firing some of our junior researchers (or at least not paying them until they threatened to take us to court), then our smaller staff would obviously find its research capabilities compromised. The problem of reduced effectiveness would probably be magnified by the fact that the junior researchers who usually did the unpleasant jobs, such as going door to door in dangerous neighborhoods to gauge public opinions on various topics of great interest, would be the primary targets of our cutbacks. On the other hand, the more highly paid senior researchers who did not really relish getting up from their desks and dealing with the grubby general public would now find themselves having to jump back into the fray. In any event, our reduced staff size would mean we have to be a little more careful in specifying our sample size. Accordingly, we would want, in an ideal world, to be able to select the smallest sample size that would still produce a given degree of confidence.

Statisticians have developed the following simple formula used to determine the minimal sample size needed for a given level of reliability:

$$n = \left[\frac{Z_{\alpha/2}\sigma}{E} \right]^2$$

Suppose we want to determine with a 99 percent degree of confidence that a random sample of test scores of students at the Dali School of Automobile Painting and

Customizing is within 5 points of the mean population value of the student body as a whole. Because of our recent staff cutbacks we would obviously be very interested in finding out how many Dali student test scores we would have to obtain in order to be 99 percent sure our sample mean was within 5 points of the mean of the population.

If we assume a standard deviation of 10, then, thanks to the heroic work of earlier statisticians who labored in obscurity for many years, we can determine how many student records we need to filch from the registrar's office (making photocopies would be out of the question due to recent funding difficulties). We would merely have to multiply 2.57 (the value for $Z_{\alpha/2}$) by the standard deviation (10) and divide that product by 5, which is our margin of error because we have defined our confidence intervals as encompassing $\mu + 5$ and $\mu - 5$ within 5 points of the mean value of the population. This process yields a quotient of 5.14: $[(2.57 \times 10)/5]^2$. We have to square 5.14, which in turn gives us 26.42. As it makes no sense to speak of a sample not expressed in whole numbers (what does it mean to refer to 0.42 of a test score?), we would then round our answer up to 27.

So what have we done? We have determined we need to get 27 randomly selected test scores from the fledgling geniuses at Dali if we want to be 99 percent certain our sample mean is within 5 points of the true population mean. Suppose we are feeling a bit guilty about the lack of precision in this example, and we want to derive a sample mean that more closely approximates the population mean. Well, we could hire some competent statisticians to carry out this calculation, or we could simply reduce the margin of error from 5 to, say, 2. How will this affect the number of test scores we must obtain for our sample in order to still maintain our 99 percent degree of confidence?

We still need to go back to the equation and substitute 2 in place of 5 for the margin of error. We would have to dust off our calculators, multiply 2.57 by 10, and divide the resulting product of 25.70 by 2 (instead of 5), which would give us 12.85. We would then square 12.85 and obtain 165.12. Again we would round 165.12 up to 166 so that we would not have to examine the profound philosophical implications of fractional test scores. Thus left with a sample size of 166, our finely honed intellects would quickly determine that 166 is much, much bigger than 27. This means that in exchange for a sample mean that is a little more accurate than our first sample (a margin of error of 2 points as opposed to a margin of error of 5 points) we would have to do a lot more work and pilfer more than six times as many folders from the registrar's office.

These results appear to illustrate the typical tradeoff between the size of the sample and the accuracy with which the sample mean will approximate the population mean. We can also look at these results another way: Our margin of error (from 5 to 2) has not been reduced by a staggering amount because we are dealing with student test scores that go from 0 to 100. But the amount of work we have to do to shrink that margin of error by 3 points will increase by more than six fold. This information could be an important consideration when we have to don our black clothes and ski masks and sneak into the registrar's office. We know that 166 files are going to be much more difficult to carry than 27 files. There is a risk that the additional files could slow us down so much that even the heaviest campus security guard could catch us. Alternatively, we might try to get the files during the day and sneak them past the ever-vigilant registrar office secretary, Miss Feldon, whose ice-cold glare could turn the bravest man into a quivering mass of jelly. We would have to come up with a clever way to get the files past her, such as stuffing them in our pants. If this was our only option then it would probably make a lot of sense to content ourselves with a sample size of 27, instead of 166, because of the potential for severe paper cuts. Needless to say, it would be very difficult to stuff 166 files into one or two pairs of pants, and we could expect a withering "Where are you going?" glare from Miss Feldon. Even 27 folders would make us look as though we had very squared-off rear ends.

What if we wanted to reduce the number of files we need to retrieve? Assuming we want to maintain that same 99 percent degree of confidence that our sample mean will approximate the population mean of all student scores, we would have no choice but to increase the margin of error. So let us assume we are in a devil-may-care mood and we decide that we will go in the other direction and increase our margin of error from 2 to 8. How will this change in our tolerance for error affect the number of files we must stuff in our pants?

To see how much we will be able to lighten the load in our trousers, we return to our equation and multiply 2.57 by 10 to get 25.70 and then divide 25.70 by 8 to obtain 3.21. We then square 3.21 to get 10.3 which is then rounded up to 11. If we do not mind allowing our sample mean to diverge by 8 points from the population mean, we can certainly stuff 11 student folders into our pants with only a few nicks and cuts. The interesting point here is that to reduce our margin of error from 8 points to 2 points, we would have to increase our sample size by more than 15-fold (11 to 166). For those of us who are not particularly anal-retentive, we may be quite happy to trade off the reduced precision in favor of a much lighter load of student folders.

This fun-filled explanation of the way to determine a given sample size does raise an interesting question regarding the supposed trade-off between the size of the sample relative to the population as a whole and the extent to which it can be expected to approximate the mean value of the population. Our equation suggests that more numerous samples do not necessarily translate into correspondingly more accurate approximations of the population mean. This is not to say that increasing the sample size will not, in the long term, result in a closer approximation of the sample mean with the population mean. The equation shows that if we are willing to accept a greater margin of error, we can greatly reduce the size of the sample.

You may be wondering whether this equation is one to use when determining sample sizes that do not involve the test scores of Dali's students. Fortunately, this equation is universal in its application. It can be used for the test scores of students of any accredited vocational school that teaches courses in automotive repair (as well as most other types of problems involving the determination of a desired sample size).

Student *t* Distributions

You will notice that in our prior sections we conveniently had the value of the standard deviation available to plug into our equation. This is one of the benefits of being the author, but it is not always indicative of the situations statisticians may face in the real world. Often the statistician will have no prior knowledge of the value of the standard deviation. If we were sloppy and incompetent statisticians, we might simply make up a number to represent the standard deviation and congratulate ourselves on a job well done, regardless of how poorly our sample mean approximated the value of the population mean. But because we take some minimal pride in our work and wish to impress any statistics groupies who happen to be hanging around the office, we may find it necessary to use what is known as the Student *t* distribution. This equation comes in very handy when statisticians either do not know the standard deviation or (more likely) are conducting an experiment in which there is not enough money or time to carry out the preliminary studies needed to determine the standard deviation. The Student t distribution was developed by William Gosset, an employee of the Guinness Brewery.[1] It was known as the Student t distribution, instead of the Gosset distribution, because Gosset's employer did not allow its employees to publish the results of their research. Gosset decided to publish his results anonymously and substituted the word "Student" for his own name. It is not known what gave Gosset this brilliant flash of insight, but it is clear he provided statisticians with a wonderful new analytical tool particularly well-suited for small samples. It was also clear that Gosset's Student t distribution could offer an economical alternative in situations where the collection of data would otherwise be too expensive or time consuming. It is used primarily when the sample has 30 or fewer events, the standard deviation is unknown, and the population is normally distributed.

To use the Student t distribution, we need to refer to a table of critical values of t that can be found in any elementary textbook on statistics. These critical values of t correspond to given areas of a normal distribution. To use this chart of critical values, we must first become familiar with the concept of degrees of freedom. This concept has nothing to do with marital status but is a mathematical device that comes into play when we are dealing with the small sample sizes handled by the Student t distribution. The degree of freedom is merely the number obtained when one reduces the number of items in the sample size by one. In other words, a sample size of 20 test scores, for example, would have 19 degrees of freedom.

What are we talking about? The nomenclature "degrees of freedom" is based on the notion that if our 20 test scores must equal a fixed number such as 500, for example, 19 of those test scores can have any number of values. However, the specification that all 20 scores must equal 500 imposes a restriction that will fall on the 20th score. In other words, we can assign a wide variety of scores to test scores 1 through 19 but we can only assign a single score to test score 20 so that we can still be certain that they will be equal to 500 as a group.

This may sound like a lot of gobbledygook. We can make the point with a simple example. Suppose we have a sample group of four test scores that will equal 70 when summed. We can assign a number of different values to each of those first three scores. Once we have specified the values of each of those first three scores, there is only one number that will still make it possible for the summed scores to equal 70. We should note we are not saying we can assign any values to the first three scores, merely that we have great discretion as to the values that can be assigned. Obviously, we could not assign a value like 275 or 821 to the scores because of the requirement that they equal 70. There would a great number of values between 0 and 70 that could be so assigned. This sample would have three degrees of freedom (three scores to which values could be freely assigned).

We would use the chart for t that would give us critical values of t based upon our specifying the area in the two tails of the probability distribution and the number of degrees of freedom of the sample (the number of values reduced by one). To return to our sample of 20 test scores, we can find the value for t by locating the vertical column in the t distribution, providing for a desired confidence interval (if we want a 95% confidence interval, then we know that the area under the tails will be equal to 0.05). We then go down to where the vertical column intersects with the horizontal column specifying 19 degrees of freedom (20 values in the sample) and obtain a *t* value of 2.093. This is the counterpart to the *z* value we dealt with earlier in this chapter.

The Student *t* distribution can be used when you have a normal, or close to normal, population distribution (symmetrical or near symmetrical bell curve). It becomes correspondingly less useful as the distribution diverges from the normal. We should point out that while the Student *t* distribution will always have something akin to a bell curve, it will vary with the differences in the number of items in the sample size. This observation makes sense because the smaller the sample, the more likely its distribution will be overrepresented by "outlying" sample values. In other words, one or two high or low scores would greatly alter the shape of the distribution curve. As the number in the sample size approaches 30, however, the Student *t* distribution will more closely approximate the standard normal distribution. In short, the Student *t* distribution, because of its smaller number of items, will have a greater degree of variability than the standard normal distribution. That variability will tend to disappear as the number of values in the sample approaches 30.

Perhaps the most basic point of this chapter is that it underscores the ways we can use statistical tools (and the comparatively limited information provided by a small sample size) to make very accurate generalizations about a population. It also serves to remind us that the process by which we select our sample is perhaps the most important part of any statistical experiment. If the sample is not representative of the population, then there will be very little useful information that even the most brilliant statistician will be able to discern about the population being studied.

Test It Yourself

All the inventions that the world contains,
Were not by reason first found out, nor brains;
But pass for theirs who had the luck to light
Upon them by mistake or oversight.
—SAMUEL BUTLER

O ne of the advantages of living in a capitalist society is that we are all free to spend our hard-earned cash on nearly any type of product we may choose regardless of whether it is moderately dangerous or simply a waste of money. Admittedly, the government sometimes interferes with our natural right to purchase the goods that make us happy, such as assault weapons and shoulder-mounted surface-to-air missiles. But, by and large, most of our acquisitive yearnings can be satisfied if we can lay our hands on enough dollars to complete the desired purchase.

In any free market economy, the consumer is supposed to be the ultimate decision-maker. After all, companies exist fundamentally to make money for their shareholders. The only way they can make money is by selling products or services to the purchasers of those goods or services. Most companies therefore have an incentive to try to determine the types of products that will appeal to the general public. For this reason, companies hire people with advanced degrees in marketing from exclusive schools to try to determine the types of products that should be manufactured. Most companies want to sell products that people will buy. There are very few companies that persist in following the marketing program embodied in the central planning methods of the former Soviet Union: The Soviet government always decided beforehand which products would be produced in a given year. This type of marketing plan presupposes that the government knows ahead of time which products the public wishes to purchase. The problem with this theory is that it does not work in practice. The Soviet government got away with it for two generations because of the vast material and human resources it marshalled, but it was unable to guarantee much more than a muscle-bound military and empty store shelves. This form of central planning helped to bankrupt the nation and in the early 1990s caused it to disintegrate into 15 independent nations.

The Soviet style of economic management is also known as the "stupid model" because it does not take into account the desires of consumers. There are not too many companies that insist on building whatever products they please regardless of whether consumers actually like their products. Needless to say, consumers do not typically spend money on products they dislike or find useless unless, of course, they are endorsed by those most revered of individuals—celebrities.

The point of this scholarly discussion of macroeconomics is that our economy is ultimately dependent on the decisions of individuals to purchase goods and services. As such, our economy is said to be demand-driven because the preferences of consumers—as evidenced by their purchases—ultimately dictate which products will be bought and sold over the long term. In a larger sense, these preferences will ultimately determine the fates of the companies themselves.

It would be overly simplistic to view the marketplace of products as being driven by the demands of consumers. Few consumers bother to clamor at a department store counter for products that do not yet exist. Indeed, it falls upon manufacturers to try to design and build products that they expect will appeal to consumers based on a whole host of considerations including the success of past products. This strategy is obviously not foolproof because even the most successful of companies have their failures. But the cost of bringing new products to market necessitates that companies try to determine beforehand whether a product has a reasonable chance of success.

What does all of this have to do with statistics? In an indirect sense, statistics pervades the entire process and can be used by companies to educate, inform, and otherwise bamboozle consumers into purchasing their products. In short, statistics is the foundation for all advertising. As such, it plays a crucial role in the formulation of marketing campaigns.

Now that we have roused ourselves from the semicomatose state in which we were beginning to fall, we can venture forth and explore the role of statistics in the modern industrial world. More specifically, we will use our newly acquired knowledge about statistics to examine critically many of the different types of claims and statements we see in daily society.

How many of us have heard commercials in which claims such as "four out of five doctors recommend" or "more people choose" or "dogs prefer" are made? These types of claims should immediately raise questions in our minds about the methods used. Take the "four out of five doctors recommend" advertisement as an example. We might be curious to know whether the survey consisted totally of five doctors, of which four favored the product and also happened to be stockholders of that company. Advertisers are not above playing games with these types of surveys. Suppose the Zippy Cola Company, long an also-ran in the carbonated beverage market, decides to undertake an innovative marketing approach and find dentists to give glowing endorsements as to the cavity-fighting properties of Zippy Cola. The challenge is to find dentists who will agree that a product laden with sugar can promote dental hygiene. The marketing department at Zippy Cola might send a crack survey team out to find a large number of dentists who would solemnly attest to the dental benefits of Zippy Cola. Unfortunately, the task would be a difficult one because very few dentists would be philosophically predisposed toward endorsing Zippy Cola. Indeed, the survey team might find itself rebuffed time and time again as dentist after dentist hung up on them.

Because of the general hostility of most dentists to a sugary carbonated beverage being advertised in this way, the survey team might decide that it would be more prudent to scale down the size of the survey and to look, in essence, for a small group of dentists who could be persuaded that Zippy Cola is the best thing for dental care since fluoridation. Instead of seeking to poll 500 or even 100 dentists, they would conduct many polls of much smaller groups of dentists, hoping to find that magical group which would accept their product as belonging right alongside fluoridated toothpaste. They might reduce their sample size to 20, then to 15, to 10, and finally, to 5. And on the 88th survey, they might actually poll five dentists, four of whom would be willing to agree that Zippy Cola is a powerful cavity-fighting product. Having finally found this unique collection of rogue dentists, the marketing division at Zippy Cola could then prepare their nationwide advertising blitz trumpeting the fact that "four out of five dentists recommend Zippy Cola to help prevent tooth decay." The ad might also give some of the reasons for Zippy Cola's cavity-fighting prowess including the fact that its sugar content is so high it leaves a thick coat of protective glucose on the surface of teeth, and that the carbonation helps to bubble away unsightly food particles and stains.

Would this be a valid survey result? No, not really. We have already seen the pitfalls associated with using samples too small to make reliable predictions about the population as a whole. This use of a single sample of five dentists out of possibly tens of thousands (and the convenient discarding of the negative results of the previous 87 surveys) would certainly suggest a certain lack of scientific integrity to the entire polling process. It would also underscore the blatant manipulation that would have to occur to enable Zippy Cola's marketers to proclaim triumphantly its dental benefits.

Similar concerns are raised with claims regarding the preferences of the population as a whole. When we hear "more people choose" a product, we should be concerned about the methods used to obtain that pronouncement. At first blush, this statement would seem to suggest that a huge sampling of the nation's population was taken and that a majority of those surveyed indicated their preference for the product advertised. However, we should be careful before jumping to conclusions about a product's popularity because of the games advertisers play.

Suppose the Dexter Bread Company decides to expand the market for its loaf products beyond its traditional customer base of French penal colonies and Dickensian orphanages. Although the Dexter Bread Company products are extremely inexpensive (due to the manufacturer's secret ingredient—they use sand instead of flour) and are popular in the tropics because of their rocklike consistency (which retards spoilage), they have never been very popular among the few consumer test groups who have tried them. As a result, the Dexter officials, alarmed at recent stagnant sales in the penal colonies and orphanages, might try to create a new image for the Dexter product line.

The Dexter brain trust might try the same type of approach as the Zippy Cola Company and hire statisticians to go throughout the country and search for persons who would sing the praises of Dexter's products. Because they would merely be seeking the simplest verification of their "more persons prefer" Dexter bread slogan, they might find it easier to obtain an acceptable survey sample because they would be dealing with the general public as a whole. Would this sample have any more validity than the Zippy sample? Well, it would depend on the way the sample results were obtained. If the Dexter statisticians took survey after survey and discarded the results of survey after survey because of the respondents' repeated inability to pledge their undying devotion to Dexter bread or even to chew it without spitting it out, then the fact that they finally did find a small group of persons who enjoyed gritty rock-hard bread with a flinty taste would not give us a great deal of confidence in the validity of the survey results. If, on the other hand, survey after survey found the vast majority of respondents dancing in the streets with loaves of Dexter bread held up to the heavens, then we would be more inclined to accept the survey as a reliable indicator of the preferences of the population as a whole.

We should also be wary of surveys that purport to express the preferences of dogs, cats, and other nonhuman respondents. Unless you have a very clever group of doggies and kitties, it is not likely you will get coherent responses to questions such as, "Did you prefer the braised liver or the sauteed shrimp?" Most cats, for example, will stare at you blankly when asked about their dining preferences, possibly trying to imagine you dead so that they can get on with their business without being further delayed. Dogs, on the other hand, are not very interested in grappling with deep philosophical questions regarding the subjectivity of sensory perceptions and would rather go fetch a stick or drink out of the toilet.

So how do pet food vendors measure the preferences of animals? One popular approach is not to feed the animal actors very much for, say, ten or twelve weeks beforehand, until they are crazed with hunger and start to imagine how their trainers might taste in a brown gravy. At that point the trainer could place a bowl of nails in front of the animal and watch it make a spirited effort to gobble it up. Not surprisingly, a hungry animal will not be inclined to be very finicky, especially when its ribs are protruding through its skin. Although an animal might ordinarily turn up its nose at the advertised product such as Smith's Goat Entrails Doggie Treats, prior deprivation might cause it to alter its preferences, if only to avoid complete starvation.

Some persons might quibble that depriving animals of food for weeks compromises the integrity of the sampling process, while others (most notably pet food vendors) would assert that lack of food clarifies the thought processes of the animals so that they do not spend so much time dilly-dallying at suppertime. The hardheaded statistician would probably be somewhat cynical that such a technique would provide a true and accurate indication of the animal's desire to consume that particular product.

How might we better measure an animal's preferences for a particular food? One way might be to line up several bowls, each containing a different brand. We would want to make sure the dogs did not see us put the food into the bowls beforehand in case they were capable of reading the brand names on the cans, thereby compromising the integrity of the "blind taste test." The food should not be too different from each other in color, texture, or consistency although certain differences in manufacture are inevitable. The bowls themselves would also need to be identical so that one bowl did not have an advantage over the others such as being topped by an alluring stuffed poodle. The bowls need to be spaced apart to prevent the smells of one brand from mixing with the smells of the other brands. Assuming that everything was on an even playing field, the only fair thing would be to release the animal from a distance and see which bowl it preferred. We would then repeat this experiment a number of times, and try the same test with other animals. If the animals truly preferred one food to another, then we would expect to see them, more often than not, veering over to their preferred food. We would want to test this preference by doing such things as switching the positions of the bowls to make sure the animals were not simply walking over to a particular bowl because it happened to be the closest or on the left or right end of the line (if the animal happened to be left- or right-pawed).

Perhaps the most important lesson to be drawn from our meandering in this chapter is that we must always be very hardheaded about concluding that a sample accurately reflects the chief characteristics of the population we are studying. In other words, we need to focus on the sample size and the degree of confidence we have in its similarity to the population as a whole. It is also important to realize it is the absolute size of the sample itself and not its size relative to that of the population as a whole that is critically important to its reliability as a representation of the population.

We can have small samples that provide accurate "snapshots" of the population as a whole. This is fortunate for statisticians because it makes it possible to gauge the sentiments of an entire city, state, or country. We can see this if we consider the dilemma of the Crusty Cheese Company, which wishes to test market its newest product, the blue cheese cheddar cake. Mind you, the folks in the corporate offices are very proud of the blue cheese cheddar cake because it represents the Crusty Cheese Company's first venture into the dessert market. They are especially excited by its unique blue swirls inlaid in bright yellow-orange cake, making it stand out among more run-of-the-mill chocolate and coconut cakes.

Now the Crusty Cheese Company wants to determine whether its product will find favor with a national audience. The initial plan, proposed by the intellectually challenged nephew of the company founder, was to mail a blue cheese cheddar cake to every man, woman, and child in the country. In this way, the nephew reasoned, the Crusty Cheese Company could truly test the public's reception to such a bold new product. Of course there were a few nay-sayers who threw a fly in the ointment of the nephew's brilliant idea, pointing out that such a direct marketing strategy would cost the company over $2 billion. For some reason, they thought this amount was too dear for a company that grossed about $20 million in sales the year before. But the nephew had pointed out that the company could borrow the money because it only had about $500,000 in its own corporate account. This strategy did not please the other officers of the company who thought there must be a better way to test the public receptivity to their new product. At this point they turned to their chief statistician and asked her advice as to how best to test market the blue cheese cheddar cake. She indicated she would rather eat rancid meat than blue cheese cheddar cake. The company officers carefully considered her remarks and then fired her, appointing her former assistant to the new corporate statistician job.

When asked the same question, the former assistant statistician, being somewhat more practical than his former boss and also having quite a talent for kissing up, would congratulate the company executives on the blue cheese cheddar cake masterstroke (while choking a piece down) and point out they do not need to inundate the entire nation with foul-smelling canisters in order to assess its probable commercial success. Indeed, the new company statistician would explain that the company merely needed to select a representative sample of the nation's population and could rely, if properly selected, on that sample as an accurate indication of the reception that could be expected with the formal unveiling of the blue cheese cheddar cake in the nation's grocery stores. How many people would be lucky enough to receive the blue cheese cheddar cakes? The new company statistician would tap his temple slowly, nodding his head as though he were pondering the question, and then suggest that cakes be sent to several thousand people who would be carefully screened beforehand to make sure they were representative of the nation's population as a whole. Cheers would erupt as the company officers realized that it was not necessary to implement the nephew's plan or even to tolerate the presence of the nephew, who could now be banished to a remote company outpost in eastern Siberia where frozen blue cheese cheddar cake might have a fighting chance.

These examples underscore the ways in which statistics can be used to simplify sampling problems, but they also illustrate the ways in which statistics can be used to manipulate data. Certainly we have all seen television reports and read newspaper and magazine articles solemnly proclaiming that a survey of some number or percentage of people had revealed their views on a particular issue of vital importance. No doubt we can relate to the broadcaster who reports to the viewing audience that a poll revealed that "38 percent of respondents spelled tomatoes with an 's' and 51 percent of respondents spelled tomatoes with an 'es' and 11 percent believed that the tomato was the modern incarnation of the Egyptian sun god Ra."

Of course the tendency of the media—whether print or visual—is to focus on the breakdown of the percentages and not to pay as much attention to either the actual number of persons interviewed or the degree of confidence. Without this information, it is virtually impossible to determine the reliability of the poll. Compounding the problem are the resource limitations and the time restrictions that are typically faced by the media. There is certainly an incentive to reduce a 4,000-person sample to 2,000 or even 1,000 persons. Most television reporters do not have a background in statistics and cannot be expected to realize that the reduction of the absolute size of the sample size can have a dramatic impact on the ultimate value of the sample. Moreover, their basic motivation is to report the news and not to analyze the reliability of their public opinion samples. Besides, there is a tendency to view public opinion as being very malleable. The temptation is to do the best one can in a very limited time because, as with the case of presidential elections, such polls will become a daily event.

All polls are not created equal. Some will describe at the outset not only the different responses received but also the method by which the respondents were selected as well as any limitations on the sampling process. They will also specify the margin of error ("plus or minus three percentage points") so that the recipient will have a much clearer idea as to the poll's reliability. National opinion polls can be extremely valid and accurate when the sampling is done with care and the questions are phrased in such a way as to minimize confusion and bias.

A properly done poll can avoid the biases that would otherwise undermine the usefulness of a large sample size. Remember, the most fundamental concern of statisticians when they compile samples is that they be representative of the population as a whole. A large sample size in absolute terms will not necessarily be very helpful if the sample itself is fundamentally biased.

We can better understand this problem if we have to deal with the ramifications of a biased sample. Suppose that Ed Smalley has—through dint of hard work and a fortuitous inheritance—assumed control of Candyworld, a leading confectionery company. Anxious to bring the company's products to new customers, Ed decides to hire a polling firm to determine whether people would eat peanut butter anchovies if they were available in leading stores. But Ed does not have any prior experience dealing with polling firms so he closes his eyes and places a finger on an open telephone book. His pick, the Beezer Polling Company, prides itself on being the fastest polling company in the business. To be able to deliver their polls quicker than their competitors, the Beezer Company does have to rely on certain arguably questionable polling techniques. In the case of Candyworld, the Beezer Company quickly determines it needs to gauge public opinion by purchasing several lists of telephone numbers. However, these lists are culled from the membership rolls of several food-tasting clubs. Now food-tasting clubs are chock-full of people who are interested in fine foods and gourmet wines and who, on average, are probably much more receptive to new types of foods such as peanut butter anchovies. However it may be a serious lapse of judgment to rely exclusively on the members of these food-tasting clubs because of that very willingness to try new foods.

The Beezer Polling Company might be content with its sampling methods, perhaps reasoning these people are the most discerning of individuals. But they are not representative of the population as a whole, which Candyworld so dearly wishes to claim as its own customers. In other words, Muffy and Biff, who frequent one of these food-tasting clubs, may be perfectly happy to give peanut butter anchovies a solid try. But Muffy and Biff would be willing to try plutonium nuggets in a cream sauce because they like to taste exotic foods. Most of the general population is not so free-spirited in its eating habits and would probably find peanut butter anchovies to be unappealing if not downright repugnant. If the Beezer pollsters excessively weighted the opinions of the food-tasting club members, their conclusions could overstate the degree to which the general public is willing to even try peanut butter anchovies. Thus, the sampling technique used by the polling firm can be critically important to the ultimate success of the product itself. If Candyworld does not have an accurate picture of the probable response of the audience to its products, then it risks expending great sums of money on a product that will not necessarily result in increased sales.

The claims of advertisers are often not subject to great scrutiny. We have seen some of the games played by advertisers who wish to create an impression in the minds of the public that may not be statistically justified. Indeed, statistics can be manipulated but it can also be used as a tool for examining the claims and statements of others. In this way it can serve as the basis by which a claim can be tested.

Advertisers often want to talk about their product or service in a way that presents it in the best possible light. We would not be surprised to hear the Nouveaux rental agency boast about the quality of its cars by claiming that "Our company has the newest fleet of cars in the industry." Or Nouveaux might even go so far as to attach a number to its claims: "Our automobiles have an average age of two years—the youngest fleet in the industry." If you were at an airport and needed to rent a car, you might base your decision as to which rental agency to patronize on such an advertising claim. After all, it would seem that the company boasting the youngest fleet of automobiles in the industry would have the most reliable automobiles in the industry. This reasoning would lead you as the prospective lessee to suppose you would be less likely to find yourself stranded on a dark country road favored by ax murderers if you decided to rent a car from the Nouveaux agency. It would seem logical to suppose the newer the car, the more likely it would be you would reach your ultimate destination.

Suppose you took the plunge with the Nouveaux agency, even though their rental counter at the baggage claim area consisted of nothing more than a gentleman who divided his time between writing up rental invoices and shining shoes. You might have greater reason for concern if the vehicle taking you to the Nouveaux lot was actually a World War II-vintage jeep mounted with a machine gun. You might be truly alarmed if most of the cars in Nouveaux's rental lot were mounted on cement blocks and seemed to date from the mid-1960s. You might ask yourself how these cars could be part of the youngest fleet in the industry as you watched a bumper fall off one of the cars while a vagrant woke up in the back seat of another. Is the claim by Nouveaux valid or fraudulent?

Deciding whether Nouveaux does have the youngest fleet in the industry would require you to test its claims by taking a sample of average ages of the vehicles and using that sample to estimate the entire fleet's age. Suppose you decided to spend a few minutes walking through the rental agency lot ascertaining the ages of the various vehicles while the attendants tried to find the tires that belonged to the 1967 Pontiac you just leased from Nouveaux. You might become suspicious about the Nouveaux claim if your wanderings through the parking lot revealed row upon row of cars built in the 1940s, 1950s, and 1960s. How might the antiquity of these cars be reconciled with the claim to have the youngest fleet in the industry?

Well, you could be having a bad dream and merely need to wake up to make everything all better. The bad dream explanation is a fairly remote possibility and you will probably not be magically transported from this lot of dilapidated vehicles no matter how many times you pinch yourself or slap yourself in the face. Indeed, it is more likely you are actually standing in a lot of very old vehicles and will remain there. Your next task will be to reconcile the claim with the reality.

Of course one of the most obvious explanations is that Nouveaux's claim relates to the age of its entire fleet and not to any specific company lot. You might have simply stumbled upon the lot where the company keeps its oldest vehicles. Or you may have a sneaking suspicion that Nouveaux's claim is simply outright fraud.

Because you have a sample of 40 or 50 cars, you might decide to calculate the average age of the sample and conduct your own statistical investigation by listing the model years of all the cars on the lot and calculating the average age of the sample. You could then use the methods discussed in previous chapters to estimate the mean age of the population of the vehicles as a whole. You would have to obtain similar estimates for the other leading rental agency fleets before you could make a conclusion about Nouveaux's claim to have the newest fleet in the industry.

Suppose Nouveaux had gone further and claimed that its fleet of automobiles were an average of 5.1 years old. This would present a different problem. We would no longer be concerned with trying to estimate the mean age of the population as that information would have been provided to us by Nouveaux. Instead we would now be using the information gleaned from our sample to test the validity of the statements by the Nouveaux corporate brass.

Now we want to use statistics to see whether the claim relating to the average age of the rental vehicles is reasonable or not based upon our own sample. Since we find ourselves with extra time on our hands due to our automobile's missing its engine as well as its tires, we can sit down and test the Nouveaux claim. Suppose we estimate the average age of all the vehicles (or what remains of them) in the rental lot where we currently find ourselves marooned is 11.1 years. We would then be presented with an apparent contradiction. How could Nouveaux have a fleet averaging 5.1 years of age when each of the vehicles in our sample is on average 6 years older? The Nouveaux officials might blame it on the new math or simply suggest that our findings are an aberration.

While we could get in a rousing bout of name calling with Nouveaux's public relations representative, we might find it simpler to return to our statistics and try to determine if the result we obtained is significantly different from that of the figure given for the Nouveaux fleet as a whole. In other words, we should be able to determine whether the difference in the mean value of our sample and the mean value of the fleet population is simply due to chance or something more sinister.

Statisticians refer to this approach as *testing a claim about a mean*. This makes sense, as we are testing a claim about the mean value of the age of the Nouveaux automobile fleet. There are other methods for testing claims about a mean but we shall confine ourselves to the traditional approach, which is perhaps the most widely used method.

The fun of testing a claim about a mean is that it requires us to stick our toe in the pool of statistical nomenclature and learn some words such as null hypothesis (which will probably not be of much use in serenading a loved one perched on a balcony). Of course, if you are very handsome or beautiful, then you can probably whisper words such as "internal combustion engine" or "dung" and spark the interest of a member of the opposite sex.

To return to the Nouveaux automobile graveyard, we would begin our investigation by stating as clearly and succinctly as possible the hypothesis we wish to examine. In this case, our hypothesis is that the mean age of all the cars in the Nouveaux rental fleet is 5.1 years. We will assume we are referring to "human years" as opposed to "dog years" even though the latter may make more sense in view of the decrepit state of the automobiles in our sample. We would note that our examination of the 35 cars in the lot revealed a sample mean value of 11.1. We would assume a standard deviation value for the population of 5.5 for the sake of convenience. Given this information, we would then ask the question as to whether the mean value of our sample (11.1) is sufficiently large to enable us to conclude within a specified degree of confidence (such as 95 percent) that the stated age of the population (the Nouveaux fleet) is wrong. If not, we might conclude the reason for the discrepancy in ages between our sample and that of the stated age of the automobile population was due merely to chance (or, in our case, bad luck).

The claim that the mean value of the age of the population of Nouveaux's automobiles is 5.1 years is what we would call the *null hypothesis,* a term that refers to a claim about some specific feature of the population—in this case, the average age of the automobile fleet. This is the claim we will directly test. To speak in the somewhat garbled syntax of statisticians, we say that the null hypothesis will either be rejected outright or that the evidence will not be adequate to permit us to reject the null hypothesis—to reject the statement that

Nouveaux's automobile population has an average age of 5.1 years. In other words, the burden of proof will be on us, and we will either be able to disprove conclusively the null hypothesis or else will not be able to marshal enough evidence to reject it. We may be able to understand this convoluted reasoning easier if our null hypothesis is that "Edgar is guilty of robbery." This would be the prosecution's hypothesis in a criminal trial. The burden of proving Edgar's guilt would fall on the prosecution. If they fail to meet that burden, then their presumption that Edgar is guilty will be rejected outright. If they do meet that burden and the presumption of guilt shifts to Edgar, then Edgar will have to provide enough evidence to disprove this presumption of guilt. If he cannot do so, then Edgar will be presumed to be guilty because the needed evidence is not sufficient to reject the null hypothesis.

In the case of Nouveaux's claim regarding its fleet's age, our evidence will either lead us to reject that statement outright or else conclude the evidence is not sufficient to warrant a rejection of that hypothesis. Given our null hypothesis, we then propose an alternative hypothesis that is necessarily true if the null hypothesis is false. This means that the null hypothesis and the alternative hypothesis will take into account all possible events (the truth or falsity of the Nouveaux claim). We also have to understand the two types of errors that occur in this framework. A Type I error occurs when we make the mistake of rejecting the null hypothesis when it is true. If we rejected the statement by the Nouveaux bureaucrats that the automobiles were 5.1 years in age when they are indeed being truthful, then we would be committing a Type I error. A Type II error, by contrast, occurs when we fail to reject the null hypothesis when it is false. If we did not reject the Nouveaux statement when it was false, we would be committing a Type II error.

We also need to include the significance level in our calculation because it is necessary to determine whether the age of the sample is "significantly" greater than that of the mean value of the population. A very common measure of significance is 5 percent (a value of 0.05), which means we will conclude our sample mean age of 11.1 is significantly greater than the 5.1 mean age of the sample if its probability of occurring is less than 0.05, or 5 percent. We can be 95 percent certain with a 5 percent significance level that the average age of our sample is significantly greater than the average age of Nouveaux's fleet. Note that 95 percent is not equal to 100 percent and that we are still dealing with probabilities and not complete certainties.

Because the number of items in our sample exceeds 30, the central limit theorem tells us that the sample mean can be estimated using a normal distribution. To calculate the standard deviation of the sample based upon the information set out here, we divide 5.5 (the standard deviation of the population) by the square root of 35 (the number of automobiles or pieces of automobiles in our parking lot), which gives us a value of about 0.93. We obtain the z value by subtracting from the mean value of our sample (11.1) the stated mean value of the population and then dividing the result (6) by 0.93, which in turn gives us a z value of 6.45. Now that this process is becoming charged with erotic tension, we move on to the table for the standard normal (z) distribution, which tells us that the critical value when we are dealing with a significance level of 0.05 corresponds to $z = 1.645$.

The point of these rankings is that we are taking our probability distribution curve and setting an upper boundary (on the right half of the distribution where the peak of the curve is split by a vertical line where the mean value of the population equals 5.1) beyond which (toward the extreme right) any value will automatically be significantly greater than that same population mean. This means we are drawing a line in the sand where the critical value appears on the curve and then seeing if the value for our sample mean falls within or lies beyond that boundary. If it falls within the boundary, we can conclude the value for our sample size is due to chance. If we find the value of our sample size falls beyond that value (to the right of the boundary on the probability distribution curve), we can similarly conclude there is too great a difference between the mean value of our sample and that of the population as a whole (given our degree of confidence) to conclude that the difference can be dismissed as pure chance. To put it bluntly, this finding would suggest that the folks at Nouveaux must have been greatly mistaken or lied about the age of their fleet. (One can only imagine a press release containing "newly discovered information" that showed that the mean age of the Nouveaux fleet as a whole was closer to 11 than 5 years of age.)

What would the verdict be in this particular case? Because the z value for our test statistic is equal to 6.45 but the critical value with a 0.05 significance value is equal to $z = 1.645$, our sample mean value of 11.1 is significantly greater than the given value of the population. This means we must reject Nouveaux's claim that their fleet population has a mean age of 5.1 years. Now remember, we are dealing with probabilities; our rejection of Nouveaux's claim is based on our assessment of those probabilities. This is not to say we can be completely certain without question that our rejection is correct. We are dealing with a high probability but we are not dealing with a certainty. As a result, we cannot be completely sure our rejection of the company's claim is justified because we are not dealing with a probability equal to 100 percent. Still, the probability we have used is sufficiently high (95 percent) to give us a great deal of confidence that the junkyard that passes for a leased car lot is not atypical of the Nouveaux fleet. We would commit a Type I error if we were falsely rejecting Nouveaux's claim but the circumstances in this case do not warrant believing we have committed such an error.

Nothing in statistics is completely certain. We could have our 95 percent certainty and still be completely wrong because we cannot account for the remaining 5 percent. Indeed, our scathing report might lead to a news conference called by Nouveaux whereby they would provide a complete inventory of every automobile they owned, revealing we had indeed stumbled into a lot that was completely unrepresentative of Nouveaux's fleet as a whole. This could be a true nightmare where, despite all of our precautions and our 95 percent degree of certainty, we could still be dead wrong. Our defense would be our proper use of statistical sampling techniques but we might still find our credibility and our competency questioned because of the fact our sample was clearly unrepresentative of the population as a whole.

Because this discussion is very fascinating for those who have not studied statistics or wintered in Siberian labor camps, we do need to reiterate that the study of statistics is the study of probabilities. We are interested in which outcomes are more or less likely than other outcomes. Accordingly, when we examine the null hypothesis and decide whether to accept it or reject it (as was the case in our Nouveaux example), we are not actually trying to prove the null hypothesis. These statistical methods for testing a claim about a mean are supposed to allow us to avoid sneaking into the Nouveaux headquarters at night and snatching all of the registration papers for every Nouveaux vehicle so that we can do our own calculation of the average age of every vehicle in the fleet. The method outlined here was intended to keep wayward statisticians out of state prison and away from hulking men named "Spike" who reek of cheap cologne because it permits us to examine a statement about a specific population without having to trot out the information about every single object in that population. After all, we can always have complete certainty in hindsight once we have reviewed the information pertaining to every single object or event in the population. Even statisticians have other things to do with their spare time, and they do not want to engage in mindless counting games when they can come to a fairly reliable conclusion regarding the information's reliability by examining a sample. In a sense, we are willing to trade off a little bit of accuracy and run the occasional risk of drawing an erroneous conclusion about the validity of the null hypothesis being studied so that we can avoid the aggravation of having to do all of the legwork necessary to be completely certain about the value of the population as a whole.

From Test Groups to Test Significance

Logic, like whiskey, loses its beneficial effect
when taken in too large quantities.
—Lord Dunsany

E very good laboratory rat knows that scientific progress, particularly in the areas of medicine and biology, is critically dependent on rigorous testing procedures by which the effectiveness of a given medication can be ascertained. This means that we have to resort to rather elaborate safeguards such as taking a large group of participants and dividing them into two groups. One group will receive the medication and the other group will get a placebo or nothing. After a certain period of time, the results of the tests will be examined. If the drug is designed to fight a certain type of cancer, for example, the researchers conducting the experiment would try to see whether the "test" group who had received the new treatment fared appreciably better than the "control" group. The most obvious question they would ask is whether the test group (which received the drug) had an appreciably lower incidence of that cancer than the control group (which did not receive the drug). If indeed there was any significant difference in the rates of recovery between the two groups of patients, then the researchers must determine whether other factors not related to the drug itself played a role. In short, the researchers must consider any possible casual factors other than the drug itself that might have affected the outcome of the experiment.

We can better appreciate the difficulty of carrying out such an experiment if we conduct our own test of the new hair restoration wonder drug "Aggranax." Because we must first show that the drug is safe before it can be approved by the federal government for sale to the public, we have no choice but to test the drug on a select group of people. We must be careful how we structure our test. In the interest of time, we might decide to go with a test group and a control group.

Who would we want to have in our two groups? First, we might want to have people who are bald. Such individuals are necessary if we want to claim that we have a hair regeneration drug. Not too many people would be very interested in a drug tested on people who have full heads of hair. It might be a good idea to have some people who are beginning to experience noticeable hair loss so that we can see how the Aggranax will affect them. One of the obvious markets for this product will be those people who are not yet bald but who are headed (no pun intended) in that direction.

Let us assume we have two groups, each having 100 people. Half of the people in each group have thinning hair and half are completely bald. We would also try to make sure that the people in each group were as equal in age as possible. This is to avoid giving one group a greater proportion of younger people (who are, on average, less further along the hair-thinning trail) and thus tainting the results of the experiment. We would then designate one group as the test group and one group as the control group. The test group would receive regular doses of the Aggranax while the control group would be given doses of something inert, such as a sugar pill.

How far we would go to try to equalize the test conditions between the test and control groups would depend on the degree to which we wanted members of the two groups to live under the same conditions while the drug was being tested. We could go so far as to house the group members and give them all the same diets. That way we would be able to control as many variables as possible relating to lifestyle choices that could otherwise muddy the results. This being the real world and not the fantasy of a research scientist, however, we would probably have to settle for sending our two groups off to their respective homes and urging them to abstain from any habits that might interfere with the experiment, including the consumption of narcotics, cigarettes, or massive quantities of alcohol. People being the creatures of habit they are, we would probably not expect our experiment to be conducted under perfectly controlled conditions.

So we would administer the drug to the test group and the inert substance to the control group and check the results at periodic intervals. With a hair restoration drug, we would want to know if it was actually causing our test subjects to grow hair. We would want to have our researchers take photographs of the participants' scalps and do follicle counts at regular intervals (e.g., weekly). We would also want to note any unusual or bizarre circumstances, such as whether Aggranax killed everyone in the test group by melting their heads or if it dissolved the lone strands that at one time covered the balding heads. In this case, we would not be able to consider our experiment an unqualified success and would, most likely, have to reformulate the Aggranax or content ourselves with selling it on home shopping television channels.

As the test progressed, our researchers would keep a log noting the changes in the subjects for both groups. If we found the control group was experiencing more hair growth than the test group, then it would obviously call into question the effectiveness of the Aggranax. We would also have cause for concern if most of the hair growth was in the nostrils, ear canals, or on the subjects' backs. No doubt the market for nostril hair-growth drugs is so small as to be economically worthless.

After a certain number of weeks had passed, we would check to see if all of our subjects were still enthusiastically participating in the experiment and keeping up with their dosages. We would hope by this time to see some differences between the two groups. We would probably also begin to see significant variations in the responses of members of each of the two groups to the dosages. In the test group, we might see some people experience significant hair growth. Others might have very minimal hair growth but exhibit side effects such as curling finger nails. Still others might show no symptoms at all. As far as the control group is concerned, we would expect fewer variations among the group members because they are not taking the drug.

One Sample, Two Sample

Our prior chapters dealt with the methods by which a sample could be used to make inferences about a population. But one cannot go through life gathering sample data simply for the thrill of calculating population parameters. It simply will not do to decline a dance with the Queen of England at a gala event because you have a sudden urge to calculate the average life of a lightbulb in Buckingham Palace. Nor can you decline the request by a mugger holding a gun to your head to hand over your valuables merely because you suddenly want to determine the probability that a left-handed child will be born in Des Moines, Iowa, on Christmas Day. Statistics must sometimes give way to the pressures of the day-to-day world.

This chapter has an element of excitement because it will deal with the comparisons of a variety of groups and ways in which information about different samples can be used to make inferences about their respective populations. Statisticians often use such comparisons to calculate the variances or means of two populations. Of course, you want to know how this information can benefit you in some noble way, such as helping you amass huge piles of cash, or making it easier for you to spend time with the beautiful people. Unfortunately, this information may not be enough to help you get your first million dollars or meet the nation's most famous soap opera diva. It can illustrate the critical dependency of our economy on the tools of statistics.

Testing variances among populations, while not as exciting as a nudist computer dating service, does have its own rewards. It is particularly useful to people who are interested in manufacturing products and marketing them more efficiently. We can see how this would work if we visited the Aimless Gun Factory in Nevada, "where quality is a state of mind." Aimless products are known around the world for their fine craftsmanship; they can send a bullet through the heart of even the stoutest of innocent bystanders. Thus, it is extremely important that the company's reputation for precision products remain untarnished. Suppose the Aimless managers are retooling the plant and have to decide which of two methods for producing gun barrels should be adopted. Because precision manufacturing is such a selling point for the Aimless product line, the managers are particularly interested in finding which production process is more consistent in meeting the various production standards. Of course the managers could always throw caution to the wind and hazard a guess as to which manufacturing process to use. That type of less-than-rigorous analysis would not impress the company's shareholders (unless the managers guessed right as to which process was better). If they were wrong, a careless selection could cause the company to lose a great deal of money and prestige. So the remaining approach would be to test the accuracy of the two production processes and, based upon the data provided by the samples, determine which process was the more accurate of the two.

We could manufacture a gross (144) of handguns using each of the two processes. Our job would be made easier if one of the processes was plainly inferior to the other, as would be the case if one process produced guns that had barrels bent like pretzels, or the guns merely came out at the end of the assembly line looking like piles of shrapnel. Assuming the other process yielded guns that could actually shoot bullets that hit their intended targets, the selection of the manufacturing process would be a fairly quick decision.

The selection process would not be so easy if both processes generally resulted in the manufacture of quality handguns. If that were the case, we would have to retrieve our statistics book and examine the consistency of these two manufacturing processes in more detail. Even if one of the processes appeared to be superior to the other, we would still have to consider whether the apparent discrepancy between the two was statistically significant.

So let us consider the wonderful world of handgun manufacture in greater detail. Our first production process involves the manufacture of fewer components. It requires more extensive monitoring of the production process because the individual pieces are larger and more detailed with numerous grooves, curves, and indentations. The second process is different because it involves a greater number of parts that are comparatively simple in design and can thus be snapped together by any minimum wage factory worker who has a degree in art history.

So we have an obvious trade-off in our two manufacturing processes. The first requires greater supervision for the manufacture of a smaller number of pieces, whereas the second requires more pieces but entails less supervision of the production process. How the relative merits of the two processes compare must await our own analysis. Let us assume we use each process to make 144 handguns each, and then begin to compile information about the two production runs. Of course, the simplest way to check the guns would be to pass them out to the assembly line workers and invite them to discharge several rounds around the factory floor. This type of testing procedure does have its drawbacks, not the least of which is the high rate of casualties and personal injury litigation. So we have to resort to more subtle, sophisticated means that can enable us to develop a quantitative standard by which we evaluate the usefulness of the two manufacturing processes.

As with any manufactured product, we are concerned with being able to produce a certain type of item in conformity with the stated specifications and with as few defects as possible. With handguns, the obvious standard (aside from whether the gun blows up in the face of the one pulling the trigger) is the accuracy of the fired bullets. We might randomly pick a sample of 10 guns from each of the two production runs to keep our sample sizes equal. After test firing each of the 20 guns five times from a fixed stand, we might then measure the accuracy of the guns by checking to see how far the bullets were located from the center of the target bull's-eye. We might find that, on the whole, the amount of variation between the two groups was negligible in that the average displacement of each bullet from the bull's-eye for each of the two groups was 2 centimeters. What would be the next step?

Even though the average displacement of every bullet for each of the two sample groups was about the same, we would want to look at the variation of each bullet from the mean. If the sample variance of the first sample was 3.15 centimeters and the sample variance from the second sample was 0.85 centimeters, then we might conclude that the guns in the first sample were subject to greater variations in their manufacturing process. After all, the larger sample variance would seem to imply a greater variation in each weapon's physical features. Even this seemingly large discrepancy could be explained away as a chance fluctuation unless we could rely on a standardized format for calculating the significance of such a difference.

When statisticians compare two samples drawn from two different populations (in our case, the two samples of handguns), they assume the two populations are normally distributed (the bell curve shape) and are independent of each other. They also compare the ratio of the variances of the two samples—the so-called F distribution—to assess the similarities of the variances of the two populations. If the ratio of the two sample variances is close to 1, then we know that the two populations have fairly similar variances. If, on the other hand, the variances of the two populations differ greatly from each other, then the value of the F distribution will diverge from the number 1.

To return to our handgun example, we would want to determine whether the differences in the two sample variances are so significant that the variances of the populations cannot be regarded as being equivalent or nearly equivalent. Having determined that the sample variances for the two samples are 3.15 and 0.85, respectively, gives us an F distribution value of 3.705. We also need to get the degrees of freedom of the two samples.

As both samples have ten values, we know that each will have a degree of freedom of 9. Armed with this information we will then saunter off to the nearest billboard featuring the F distribution and find the value for 9 degrees of freedom in both the numerator and the denominator where we are dealing with an α value of 0.025 in the right tail of our probability distribution. The F distribution for 9 degrees of freedom in both the numerator and denominator is equal to 4.0260. No doubt you are seeing the same type of events we saw in previous chapters because you know these handy tables will once again provide us with the "thumbs up" or "thumbs down" evaluation of whether the two variances can be regarded as equivalent. We will find that the value for F is equal to 4.0260 but the test statistic is 3.705. This means the test statistic falls outside the critical region of the bell curve. As a result, we do not have enough evidence to reject the claim that the two variances are equal. In other words, the difference in the variances of the two samples is not so great that we can assume that the variances of the populations themselves are not equivalent.

Statistics is, first and foremost, the study of probabilities. Our use of samples as opposed to entire populations precludes us from being able to say with complete certainty that a particular hypothesis should be accepted or rejected. Instead we base our conclusion on probabilities that provide us with a level of confidence that a statement is true or, alternatively, that a statement should be rejected. Statistics will not be completely satisfactory to those who are content with nothing less than complete certainty. Such certainty defeats the purpose of analyzing a sample to draw conclusions about the population from which it is drawn. In the case of our handgun sample, those who want perfection will not accept the results of our sample as being conclusive because they will want to test every single handgun's accuracy from both production processes. This type of testing would, of course, defeat the whole purpose of using statistical analyses because it would then simply boil down to examining every single gun in the production process and comparing the aggregate accuracies of both groups of 144 guns before making a decision as to which process should be used. Proper sampling, by contrast, can provide us with a near certain (such as 90 percent or 95 percent) level of confidence without requiring us to test every single gun in the process. Statistical sampling is particularly useful when the testing process will result in the destruction of the object being tested.

Suppose you operated a gunpowder manufacturing plant. The only way you could test the product would be to pour a pile of gunpowder onto the ground, toss a match on it, and run away very quickly. Of course this testing process would cause the gunpowder to explode thereby destroying the sample. So it would be very self-defeating to conduct 10,000 tests each day of all of the gunpowder we produced because there would obviously be nothing left to sell to paying customers. Now this scheme might not bother pyromaniacs and those ne'er-do-wells who like to see things blow up, but it would surely not help us develop a profitable business. After all, we could strive for that 100 percent certainty that we are producing premium-grade gunpowder by testing every single batch. But our satisfaction at being such a first-rate producer of gunpowder would quickly dissipate once we realized there was no product left over to sell.

Significance and Insignificance

Statisticians spend much of their time debating whether the discrepancies between two sample groups are significant or not. We spent some time earlier discussing the distinction between discrepancies that are merely due to chance and those that are statistically significant. What is statistically significant for a theoretical investigation may be meaningless in the everyday world.

No, this is not an admission that everything we have discussed in this book has no practical significance. Indeed, it should be clear by this point that statistics is, by its very nature, practical in its orientation and value. There is certainly a distinction to be drawn between results that have importance as a statistical experiment and results that have importance to us as individuals. Suppose our good friends, the researchers at the Broccoli Institute, a trade organization representing broccoli growers throughout the country, discovered that human beings who lived solely on broccoli and oat bran lived, on average, three months longer than those people who lived on a more conventional diet boasting all of the major food groups. Assuming that the proper experimental controls had been implemented, this finding would be of great interest. Indeed, we would probably soon see numerous articles in the newspapers and health segments on television talking about the wonders of the "broccoli regimen." No doubt more than a few people would publish books about the broccoli lifestyle. We might also expect companies to get into the act, marketing everything from broccoli in pill form to "I love broccoli" T-shirts. There might even be a resolution passed by the U.S. Congress to make broccoli the national vegetable. A ticker-tape parade might travel through New York City with a person dressed as a six-foot stalk of broccoli seated in a convertible, waving to the cheering crowds.

All of this hoopla would not be unexpected and would probably continue unabated until someone, such as a small child (as with the emperor's new clothes), innocently asked his or her parents why anyone would want to live on a diet solely of broccoli and oat bran. With this point made, it would only be a matter of time before people started asking themselves why in the world they would want to have anything to do with a broccoli-based diet. Sales of broccoli collectibles would plummet. Stores would find themselves with huge inventories of unsalable broccoli-colored suits, skirts, and sweaters. Children would no longer want to dress up as broccoli on Halloween. Dogs would turn their noses up at broccoli chew toys. Candidates running for Congress might attack their incumbent opponents for voting for the broccoli resolution. All over the country people would be turning away from broccoli and seeking answers somewhere else, possibly in the meat and dairy sections of the supermarket.

The point of this digression is that a diet consisting solely of broccoli and oat bran might result in an increase in life expectancy that is significant from a statistical point of view. By "significant," we do not mean that three months of increased life span is significant; we are instead referring to a determination that the result (the conclusion showing that one diet leads to a longer expected life span than the other) is statistically meaningful. However, the blandness of such a diet would probably not be of great appeal to many people, particularly to those who had in a short-lived fit of religious hysteria given all their worldly possessions to various vegetable cults. Even though the diet might promise a little more time on earth, it is a good bet that most people would probably not be impressed enough with an additional three months of life, particularly when that three months would come at the price of 20, 30, 40, or even 50 years of culinary monotony. By that point, many individuals might be in such a bad mood from having been deprived of meat, cakes, candy, and all the other tasty things that make life a wondrous carousel. The extra three months of life might seem more like three additional months in vegetable hell and might not be something about which they would be particularly enthused. Even though the regimen might promise, all other things being equal, a modestly increased life span, the expected reward might simply not be great enough to excite very many people. As a result, we would find ourselves in a situation where the theoretical significance of the experiment would not necessarily translate into any meaningful practical significance. Most people would simply not find the benefit of the broccoli and oat bran regimen to be worth the cost of the ongoing dietary deprivation.

Nonquantitative Statistics

As this is a profound book, intended to inspire its readers to think deep thoughts, we must step back for a moment and consider the extent to which statistics has become an integral part of our modern society. It is an essential tool of any company seeking to rationalize its production processes so that the greatest number of goods can be manufactured with the fewest defects. It is also very useful to politicians, as we have seen, who wish to gauge public support for various ideas or for their own electoral ventures. Nearly all government agencies use statistics to analyze the consequences of various policies and programs.

Because all governments must operate with limited resources, it is extremely important that they be able to marshal certain mathematical tools to assess the effectiveness of a particular policy. After all, the public arena lacks a true marketplace whereby buyers and sellers can in unison decide how much of a good should be bought and sold. Consequently, the government must hire statisticians to determine the quality of a program and the degree to which its benefits can be measured and, ultimately, justified.

Suppose the government decides that it wishes to measure the happiness of the general population. Now happiness is not a concept that is quantitative. You do not typically wander up to strangers and say, "You must have about 30 pounds of happiness around your waist." Such remarks would probably be misconstrued as an insult. Instead the perception that someone is happy is dependent on our perceptions of the individual's mood. Some tell-tale signs of happiness include skipping along the sidewalk, having a dopey smile, and, in some cases, voting a straight Republican ticket in the general elections.

We might want to send our statisticians out with questionnaires laden with queries designed to measure the respondents' general attitudes toward such things as themselves, their careers, their families, and friends. We would be dealing with a nonqualitative standard (degree of happiness) as opposed to a specific mathematical quantity such as average household income or property prices, so we would have to devise a ranking system to measure the happiness of individuals.

This sounds like a good plan in principle but the execution can be somewhat messy because of the slippery nature of such subjective concepts. We could hang out on a streetcorner, clipboard in hand, and stop passers-by to ask them whether they are happy or not. ("Score one for a happy person and zero for a sad person.") We would not have a separate scoring system for pathologically violent people or other deviants who might wish to participate in our poll. As the day wore on, we would continue putting an "X" in the appropriate category so that we could keep track of how many people said they were happy and how many were not. At the end of the day, having weathered the indignity of numerous puddles being splashed on our suits by passing cars and the occasional "bean counter" insult, we would review our results and find that we had scored 148 "happy" and 113 "unhappy."

Would we be very comfortable with the outcome of our poll? Would it make us confident that we would be up for "Statistician of the Year" honors? Would we even be able to say we felt that our respondents had answered the questions accurately and honestly? Given the very limited nature of our inquiry, we would probably not be able to answer any of these questions.

And what about the value of the survey itself? How many blue-chip corporations would be lining up at our doors to purchase our survey results so that they might better refine their own marketing efforts? How many advertising agencies would want to review our data so that they might be able to target those two massive advertising audiences—happy people and unhappy people? Sadly, no one would likely give us more than a condescendingly pathetic shake of the head because our happy/unhappy polling results would not strike them as something being on the leading edge of statistical research techniques.

How might we make our happy/unhappy poll a little more useful? First, we have to decide why we are collecting this information. Are we trying to measure the general contentment of the population, or are we trying to find out something more specific such as whether the population's feelings of depression would guarantee the success of Zippy Cola's new amphetamine-laced Rocket Cola? Assuming we wish to exploit the results of our survey for blatant commercial gain, we would want to develop a methodology that would enable us to determine how people define the term "happiness." In short, what do people mean when they say they are happy? For some, happiness is finding a shiny nickel underfoot when they thought they had stepped in a wad of chewing gum. For others, happiness is more elusive, ranging from the discovery by the disgruntled wife that her marriage license is void to the chance encounter by a tourist in New York City with that rarest of wonders—a working public telephone.

Some statisticians might try to skin the problem of defining happiness by resorting to helpful classifications. If we asked a person whether she was happy with her job, for example, she could give any number of responses such as "extremely happy to the point of delirium," "happy but feeling vaguely dissatisfied with the human condition," "somewhat happy but beginning to take extra-long lunches," or "extremely unhappy and now planning to purchase an assault weapon for the upcoming company picnic." These responses suggest a sliding scale of happiness or grades of happiness but they do not really provide us with a quantitative standard that can be replicated among different individuals. After all, one person might become so distraught after being admonished by his manager for parking in the wrong parking space that he would wire himself with explosives for the upcoming sales conference. Another person, by contrast, might have a more sedate attitude toward such things and would view a tongue-lashing as a pleasant experience to be savored because it allowed him to be the center of attention of the person he most wanted to impress—his manager.

Another scheme might be to try to assess the things that are most important to peoples' sense of happiness and allow them to rank the relative importance of a given set of things that make them happy. We could ask our respondents to tell us how they would spend a fixed number of dollars to purchase those things that give them the most pleasure. We would probably not be surprised to find our respondents picking such things as new homes, new cars, new spouses, new pets, new clothes, new personalities, and new careers. Of course not every selection would be practical but it would give us some idea of the preferences of the respondents. If, for example, most of our respondents chose a home, then we might be justified in concluding that money can buy happiness insofar as it relates to the acquisition of a two-story colonial in the suburbs. If every respondent listed something different, whether it was a hairstyle that resembled a sunflower or an iguana with a striped tail or even the head of a statistician suitable for hanging over the mantlepiece, then we would have no real way to rank the preferences of our respondents because of the absence of any underlying pattern.

Another problem we might encounter with a topic that is arguably abstract is that the answers provided by our respondents would be no more concrete and measurable than the questions. Whereas some people might find happiness in owning a new motorcycle or a gaudy diamond necklace, others might find happiness in "the rising of the sun" or "the smell in the air after a thunderstorm" or "the lapping of the tides on a sandy shore." Statisticians hate these types of "contented" individuals because they further complicate an already difficult mathematical problem by giving nonquantitative responses to nonquantitative questions. For these reasons, then, statisticians will generally prefer to try to construct their experiments in some type of mathematical framework in which only quantitative answers may be given. For the modern corporate marketing executive, the best types of surveys are ones that reveal the characteristics of the target audience and their preferences as consumers.

Suppose we are the officers of the Vulcan Match Company, and we are in the process of designing an innovative new match that we believe will revolutionize the match industry. Our match design team has reversed the traditional match design of a wooden stick tipped at one end with a chemical coating that would burst into flame when struck against a rough surface. Instead our research team proposed that almost the entire length of the match be coated with the same chemical coating, leaving just a small bit of uncovered wood at one end. After seeing the difficulty of holding the uncovered wooden end in one's fingers, we decided to take the design change one step further so that the match would not even have to be struck to be lit. Suppose our team coated the prototype matches with a special chemical that caused the match to burst into flame once it was exposed to oxygen. This new design quirk would also necessitate the design of a special box in which the matches would be kept sealed until the owner pressed a small button. A match would then slide through a small hole on one end and burst into flame.

Quite impressed with our ability to so greatly complicate the traditional match and matchbox, we would then want to hire a statistician to do a marketing survey. After all, we would want to see who would be interested in this revolutionary new match and match-box called the "Sentinel." We would direct our corporate statistician, Ms. Klepp, to help us identify the audience that would be most interested in this type of product and also to see whether they would prefer that any additional functions be included with this product. At the very least, we would want to be certain there is an actual market for this product so that we do not go ahead with a full-blown production scheme only to find that nobody wants to purchase the Sentinel.

Having been given her marching orders, Ms. Klepp and her team of assistants would then begin reviewing both the marketing campaigns used by our company as well as our past competitors. This would at least provide them with a basic knowledge of the existing market for conventional matches. They would find that smokers offer the biggest single market for matches, followed by candlestick testers and arsonists. There might also be some consumers who merely liked the feeling of carrying a box of a thousand matches around in their pocket or purse so they would know they could light a neighbor's cigarette on request or have a source of light should they choose to venture into the local sewers and trap rats.

Ms. Klepp would also want to build upon this existing marketing information to see if there would be any new markets for the Sentinel. Since it would be much more costly to manufacture than its wooden and paper counterparts, the Sentinel might not be a mass market product at all but, instead, an expensive specialty item that would be sold only at very pricey smoking and gadget stores.

Once Ms. Klepp felt reasonably confident that the basic markets had been identified, she would then try to generate a representative sample of respondents who could be asked questions about the possible attractiveness of the Sentinel product. This sample would depend in large part on the existing and near-future marketing plans of Vulcan. Our existing marketing channels would determine who receives information about our products and frequents the places where our products are purchased. Ms. Klepp might decide to acquire mailing lists from the stores that carry other Vulcan products and query people directly through the mail and over the telephone. Alternatively, she might suggest that Vulcan use a more "on-the-street" approach and set up testers at a number of selected sites so that people would have an opportunity to actually see how the Sentinel works.

After some discussion, Vulcan might authorize Ms. Klepp to employ a two-tiered approach, involving both the mailing of marketing surveys to several thousand people randomly selected from both the Vulcan mailing lists as well as mailing lists acquired from some of the stores carrying Vulcan products. At the same time, Ms. Klepp would be given the authority to hire several dozen assistants who would be placed at tables at specific locations, such as malls and grocery stores, to let people see the Sentinel and try it out for themselves. Their reactions to the Sentinel would be duly recorded as well as their thoughts about its desirability and any additional improvements it needed.

Once all of the mass of survey data was received and correlated into a digestible format, it would be up to Ms. Klepp to summarize the findings and make her recommendations to Vulcan. If the initial responses were very favorable, then Vulcan might go ahead and direct one of its factories to begin production of the Sentinel. Similarly, a very unfavorable result would probably convince Vulcan to scrap the Sentinel project altogether. What if the results fell somewhere in the middle, indicating a mixed response to the Sentinel? Vulcan might still choose to go full steam ahead or it might choose a more limited production run and see how the product fared in a few selected stores. On the other hand, it might decide that an inconclusive response was too risky and still scrap the idea. One other option might be for Vulcan to do additional marketing surveys, perhaps using an outside organization so as to get another perspective.

One unsettling fact about marketing surveys is that they are inexact. How a question is phrased can greatly affect the type of response elicited from the respondent. Indeed, the demeanor of the questioner can make all the difference in how the respondent perceives the product. Certainly marketing research is a discipline that can be carried out in a variety of ways with significant potential for success or failure based on things that have nothing to do with the merit of the product itself. It is the skillful market researchers who are able to determine when people are reacting to the product itself as opposed to something else that is irrelevant to the product's use or value.

Correlation and Regression

For her own breakfast she'll project a scheme
Nor take her tea without a stratagem.
—EDWARD YOUNG

W hen we consider the true astounding achievements of modern civilization, we think of the
invention of the printing press, the discovery of Newto-
nian physics, the Industrial Revolution, Clerk's discovery of electromagnetism, Mendel's invention of genetics,
Darwin's theory of evolution, Einstein's theory of relativity, the discovery of quantum mechanics, and the
concepts of correlation and regression. Well, perhaps
correlation and regression are not quite in the same
league as these other pinnacles of human achievement,
but they are still among the most important tools available to statisticians. They help us to identify patterns
amidst large quantities of data from which general
conclusions about that data can be made.

Now many of you would no doubt like to understand better the concepts of correlation and regression
because they are very impressive-sounding words with
a total of seven syllables. But there are other reasons for

using these words than simply polite chit-chat with fellow straphangers on the subway. After all, they represent concepts that are of great value in helping statisticians organize data. You can be the judge of that statement once we examine these concepts in greater detail.

Our master plan requires us to proceed alphabetically, so we shall begin with the concept of correlation. What is correlation? Simply put, it refers to the extent one type of variable corresponds or relates to another type of variable. In other words, we are interested in whether there is any correlation between two variables when we try to determine whether a change in one variable will be accompanied by a similar change in the other. Statisticians are also interested in whether the relationship is positive or negative. Here we are not concerned with the "good karma" or the "bad karma" of the variables themselves but with the patterns they display. By patterns, we mean the changes that occur among two variables and the directions in which those changes manifest. If one variable increases in value, for example, we would look for increases or decreases in the value of another variable to determine whether there is a significant degree of correlation.

To take a common example, many statisticians have tried to determine whether changes in family income directly correlate with the academic performances exhibited by the children of those families. The natural tendency would be to assume the children from wealthier families would do better in school because they could afford such things as new books, calculators, and tutors. A careful statistician might also try to see if there was any inverse relationship that would indicate whether increases in family income correlated with diminished academic performances (perhaps because the children were already planning to live off their expected inheritances). Moreover, that statistician would want to evaluate the intensity of degree of correlation of the variables. The changes associated between two variables might be very intense, or they might be negligible and thus ignored.

We can better understand the concept of correlation by the use of a graph. Statisticians use a two-axis (x-horizontal and y-vertical) diagram called a scattering diagram to examine the correlation between different types of variables. Suppose we are interested in finding out the relationship between the number of cans of Avalanche Beer ("America's first 200 proof beverage") consumed by individuals and the maximum possible distance these individuals are able to drive before running into another car, pedestrian, or stationary object such as a mountain or a tree. This data might be extremely important to the manufacturer of Avalanche Beer if it is being featured in many unflattering news reports questioning the safety of its product. As a result, Avalanche might adopt the approach that the best defense to such charges is to mount a good offense— namely, to assert that Avalanche Beer is a vitamin-enriched health food. This type of advertising would certainly catch many people off-guard and be admirable for its complete lack of reality. It would also necessitate that Avalanche develop some type of research to back its claims. Certainly Avalanche could drop some vitamins in each can of beer during processing. But vitamins would probably not be enough to counter the unflattering images associated with the product. Avalanche might want to go to the heart of the matter and prove, if at all possible, that the consumption of its beer would actually improve one's driving ability.

So Avalanche's crack research staff would go to work. They would first set up a test course and select two groups of test drivers, one of which would be allowed to consume large quantities of Avalanche beer. Once this group of test drivers made it over to their automobiles (or were tossed into the front seats by the always helpful staff), the actual testing of this provocative idea (that Avalanche beer would improve driving skills, especially as one passed the legal limit for the consumption of alcohol) could take place. Suppose the first drunken driver lost control of his car and careened into the retaining wall. We could view that result as an indication of things to come, or we could remember who is underwriting this entire project and dismiss it as an aberration. Suppose the second driver managed to make it down the first curve before turning the car too sharply and flipping over. We might wonder about the odds of getting two incompetent test drivers who could not even drive down the initial stretch of the test course. If the third driver plowed into a pit crew, then we might begin to wonder if there was something to this radical idea of excessive alcohol consumption leading to more accidents. But we would then remember the sponsor of our research and resolve to keep a more open mind as to other possible factors such as pebbles on the pavement that might force a vehicle to skid out of control or even the possibility that sinister beings from another world were shooting a death ray and deliberately causing the cars to crash.

As the carnage of our test group continued, our statisticians would begin recording the data so that it could later be put into graphical form. Our recorder might write that driver #1 drove 24 feet, driver #2 drove 67 feet, driver #3 drove 86 feet, and so on. We would also want to include the amount of beer consumed by each driver prior to beginning his or her drive. So the recorder would note that driver #1 had two cans of beer, driver #2 had one can of beer, driver #3 had one can of beer, and so on. Once all of the Avalanche drivers had finished crashing their cars and the fiery debris had been cleared from the track, then the statisticians could begin transferring the data to a scatter diagram to see how the number of Avalanche beers relates to the distance an Avalanche beer consumer can drive before crashing or colliding. The horizontal (x) axis could be scaled to represent the number of beers consumed, whereas the vertical (y) axis could represent the distance driven prior to the collision.

Because Avalanche wishes to promote the idea that its beer will lead to more clearheaded thinking, it would hope that the data would show a positive relationship between the number of beers consumed and the distances the drivers could operate their vehicles before crashing. Avalanche would like to see that driver #8, who consumed seven cans of Avalanche beer, drove around the test course without incident instead of passing out before he could climb into the car. It would also prefer that driver #5 who guzzled a record 12 cans of beers easily handled the hairpin turns at the far end of the course instead of flipping his car into a ditch after trying to stand up on the hood of his moving vehicle to "kiss the sky." The net result would probably not be a great surprise to those of us who have tried to drive a vehicle under the best of circumstances with all of our faculties intact. It might dishearten the Avalanche executives to see there was a very strong negative (inverse) correlation between the number of beers consumed and the distance driven. In other words, the more beers consumed, the shorter the distance traveled.

To show this relationship, we would transfer the collected data to this two-dimensional graph. In the case of driver #1, we would go to the "2" on the horizontal axis (because that was the number of beers he consumed) and then move upward vertically until reaching a point where 24 feet would be marked on the vertical axis. At this junction we would place a dot. We would then repeat the process for the remaining drivers. As more and more dots were added, we would begin to see a pattern emerge on the graph. The greater the number of measurements (dots), the more detailed the pattern.

Granted, mapping out statistical data on a chart is not the most entertaining way to spend the day. But as the dots are added to the graph, the general orientation (slope) of the pattern will become clearer. The slope of this line will give us some idea of the degree of connectedness between the number of beers consumed and the amount of distance traveled. If the line was perfectly flat (horizontal) so that it extended outward from the vertical axis of our graph without sloping upward or downward, we could conclude all our drivers traveled the same distance regardless of the number of beers they consumed. Similarly, a line that extended upward from the horizontal axis would lead us to conclude that all our drivers consumed the same quantity of beer but were able to travel vastly different distances.

The completely vertical or horizontal lines would not be a likely outcome given the data provided by our drivers. Indeed, we would expect a negatively sloped line of the type shown in Figure 5 that would show that the greater the quantity of beer consumed, the shorter the distance traveled by the test driver. As the number of beers consumed increases, the tendency of the test drivers to run into objects increases because the more intoxicated drivers are unable to drive as far as their less intoxicated colleagues. These results would also have to be compared to those obtained by a control group whose drivers consumed nonalcoholic beer to be meaningful.

Had the fondest hopes of the Avalanche executives been realized and had our drivers been able to travel progressively further distances by consuming evergreater quantities of Avalanche beer, the slope of the line would have been positive, as shown in Figure 6.

This simple illustration suggests, at least on the surface, that there is a linear relationship between beer

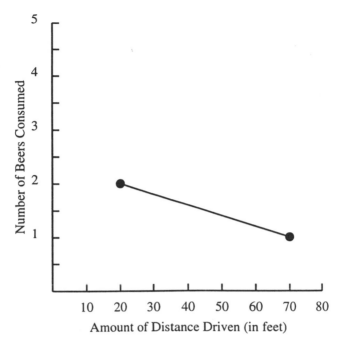

Figure 5. Negative relationship between number of beers consumed and amount of distance driven.

consumption and a propensity to drive into objects and, unfortunately, into people, which is not amusing but instead a recurring tragedy.

We have to be careful when looking at these types of scatter diagrams. It is often tempting to draw conclusions from them that may not be as evident as we might think. Certainly we want to be able to measure the intensity of the relationship between the variables. Perhaps the simplest way is to focus on the patterns of the dots themselves. This is not necessarily the most mathematically satisfactory way to express a linear correlation, but it is fairly straightforward and can be easily understood.

If we were to go to a symposium for statisticians, we would hear many terms being bandied about over

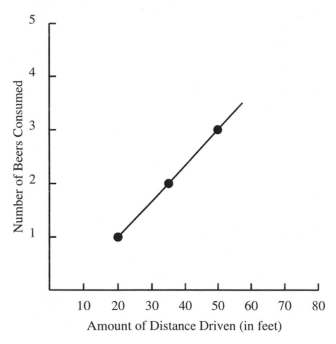

Figure 6. Positive (fanciful) relationship between number of beers consumed and amount of distance driven.

tea such as "positive correlation," "negative correlation," and, on occasion, "unrestricted corporate grants." Statisticians are immensely interested in the concept of correlation because it makes it possible to see whether there are certain fundamental relationships between variables. This is not to say that a finding of positive correlation between two variables (such as the number of beers consumed and the distance driven by the drinker before running into a heavy object) is the be all and end all of the statistician's mission. Indeed, it is often as significant to find a negative correlation between two variables or even no correlation at all.

Let us take the idea of negative correlation first. By way of comparison, we say there is a positive correlation between two variables when increases in the values of one of them are accompanied by increases in the values of the other variable. A simple illustration is the positive correlation between the amount of onions and garlic one consumes for lunch and the distance which other passers-by travel to circle around you on a busy sidewalk. If you eat one onion, most people will probably not need to give you much breathing space. If you eat several onions and maybe two or three cloves of garlic, you will probably find that people will begin to cross the street to get out of your way, run inside a building for fresh air, or even walk around entire city blocks. This example might show us that the consumption of more and more onions and garlic would have a positive correlation with the distances people were willing to travel to avoid contact with us.

What if we were in a closed area such as an elevator? Our search for a positive correlation would have to be adjusted somewhat. We could still have our test subjects increase their consumption of onions and garlic to excruciatingly high levels. However, the other variable would have to be altered from distances traveled to avoid contact to perhaps the number of attempts by the captive audience to break down the doors of the elevator or perhaps climb through the ceiling. Regardless of the factual circumstances, the key here would be to find one variable that increases or decreases directly with similar changes in another variable.

Having reinforced the concept of positive correlation, it is not a great leap to the concept of negative correlation. Here, we are looking for an inverse relationship between two variables where increases or decreases in the value of one variable are accompanied by decreases or increases in the value of the other variable. A good example of an inverse relationship is the amount of horsepower in an automobile engine and the number of miles that engine will power an automobile on a single gallon of gas. As with our other examples, this may not be a perfect correlation. This would be the case if a change in one unit of the first variable invariably led to a one unit change in the other variable. If we know anything about the combustion systems of automobile engines, we will have a sneaking suspicion that the more powerful, more complex engines will require more fuel to power a car a certain distance. Of course, we will have to assume that all other factors are equal, such as the conditions in which the engine is tested and the weight of the vehicle. We will expect, on the whole, to find the more powerful engine to be a thirstier gas guzzler.

To test this theory, we would secure several engines of differing horsepowers. We might want to make sure they were made by the same manufacturer and in the same condition so that we did not skew our test results. We could then gather our crack team of mechanics and have them install each engine into an automobile chassis obtained at a very reasonable rate from a local chop shop. We would make sure that exactly one gallon of gas was placed in each of the engines prior to ignition. We would install the 79-horsepower Johnson Butterfly—an engine more suitable for powering rickshaws—and then drive the automobile around the test track until it ran out of gas. By totaling the distance traveled, we would then be able to obtain a miles per gallon total for that engine. A more reliable measure might be to put 10 or 20 gallons of gasoline in each engine and divide the resulting total number of miles traveled by the number of gallons of fuel. This average mileage would help us take into account the efficiency of the engine, in terms of gas consumption, as it reached an optimal state of performance.

After testing the Johnson Butterfly, we would move to more powerful engines such as the Johnson Hedgehog, a squatty 110 horsepower engine used in police assault vehicles and amphibious lawn tractors. We would repeat the same test with the same amount of gas and measure the distance driven. Next we would install the Johnson Wildcat, a purring 220-horsepower engine used in many high-performance compact cars. This would be followed by the Johnson Buffalo, a 440-horsepower behemoth found in many drag racing vehicles. (It won its name by sounding like a herd of buffalo and emitting exhausts reminiscent of the stables at a slaughterhouse.) Finally, we would install the ultimate Johnson engine, the 780-horsepower Tyrannosaurus Rex, a testosterone-inspired monster so large the front end of the automobile chassis had to be removed in order to mount it onto the frame. Aside from the fact the bottom of the front end of the automobile tended to scrape the highway, we would no doubt be very impressed with the near-supersonic rate of acceleration (which caused our test driver's lips to wrap around his head) and the sheer thrill of having so much power at our disposal. Whether we would be as impressed with its fuel efficiency would be another question that would have to wait until our researchers finished tabulating the results of our test drives.

So what would we find out once we began plotting the data points on a two-axis graph? We would first have to decide how to depict the information so we could plot the amount of horsepower on the vertical axis and the distance traveled on the horizontal axis. The Johnson Butterfly, being the lightweight of the group, might run 53 miles on a single gallon of gas, whereas the Johnson Hedgehog would chug along for 48 miles on a single gallon of gas. This would appear to suggest an inverse or negative correlation between the amount of horsepower and the distance traveled on a single gallon of gas. But two plots do not make a correlation, and we would have to press forward before we could make the kind of sweeping generalization so dear to most television infomercials.

So we would continue to plot the remaining points detailing the number of miles traveled by each engine. All in all, we would probably find there continued to be a negative correlation between the power of the engine and the miles traveled. This conclusion would be expected because we know that there is a tradeoff between an engine's efficiency and its horsepower. However, this tradeoff may not be precise. The plotted points will rarely fall in a perfect line proceeding upward at a 45 degree angle from the origin (in the case of a positive correlation and the reverse angle in the case of a negative correlation). Instead some of the points will be scattered above this 45 degree line, and others will be scattered below this line. The further these lines are located above or below this line, the weaker the correlation. The greater scattering will indicate a lower degree of correspondence in the relationship between horsepower and mileage.

This inexact inverse relationship is shown by the scattering of the points around our 45 degree line. We might want to hazard a guess as to why this relationship is not a perfect correlation. Indeed, it may be that the smallest Johnson engine is simply too underpowered to move the chassis very well so that the expected savings in gas mileage does not materialize. As a result, there may be little difference in mileage between the Johnson Butterfly, Johnson Hedgehog, and Johnson Wildcat. Indeed, the Hedgehog and Wildcat may show very minor variations even though their horsepowers differ. The smaller engines may have to work more to power the same chassis, and this extra effort could reduce their operating efficiencies.

We would expect the Buffalo and Tyrannosaurus Rex engines to confirm our expectation that the more powerful engines would have reduced operating efficiencies. Indeed, the Buffalo engine would roar like a lion and throw off huge plumes of smoke, creating a pyrotechnic display to rival the Fourth of July. Of course there would be a price exacted in terms of consumed fuel and charred pit crew members, but some people might find the price to be acceptable. The reduction in mileage would likely be even more pronounced with the Tyrannosaurus Rex, an engine whose mileage is measured in terms of gallons per mile instead of miles per gallon. This top-of-the-line Johnson product would make the ground shake and cause fire to fall from the sky every time its driver pushed the accelerator to the floor. Unfortunately, it would also run out of gas before it moved very far. So the bigger Johnson engines, despite their very impressive noises and visual features, would probably only confirm our suspicion that the higher horsepower will result in lower mileage.

Our plotted diagram will show a negatively sloped line reflecting the negative correlation between mileage and engine horsepower. The degree of correlation may not be perfect, due to the factors noted earlier, but we will expect overall to see a negatively sloped line. Whether the degree of correlation can be improved may well depend on our obtaining additional measurements of engine efficiencies. For example, we could run each Johnson engine through a series of tests and plot their results. As more and more tests are conducted, we may find that the plotted lines will begin to merge into something approaching a solid smear of ink. Indeed, 50 test runs (10 for each Johnson engine) will likely reveal a stronger correlation and a more defined graph. Additional tests may simply reinforce that finding.

Although this plotting of points on a diagram seems like a fairly effective way to highlight a positive or negative correlation between two variables, it does not sit well with statisticians because it cannot be easily expressed in a mathematical form. Moreover, it relies on our ability to accurately place points on a two-dimensional graph even though we may have to deal with very large numbers. This desire for greater quantifiable precision coupled with the fear of having to create graphs the size of a football field to handle large correlations prompted the development of a statistic called the linear correlation coefficient. The mathematics of this statistic is somewhat complicated, but it does provide a sample statistic that will give a mathematical expression for the degree of correlation.

Cause and Effect

When we see a positive or negative correlation between two variables reflected on a two-dimensional graph, we can see there is some type of relationship between the two variables. In the tests involving the Johnson engines, for example, we determined there was, by and large, a negative correlation between the amount of horsepower in an engine and the number of miles the engine could travel on a gallon of gas. Although this correlation might not be perfect or of equal intensities throughout the diagram, it will reflect a general tendency among the variable designating miles traveled on a gallon of gas to decline with each successive increase in the engine's horsepower. The fact there is this linear relationship does not mean there is a cause-and-effect relationship. In other words, the variations in one variable do not actually cause variations in the other variable. The changes in one variable are merely associated with the changes in the other variable. At best, the linear relationship simply describes the amount of change occurring in one variable as the other variable is altered.

Suppose the Muscleworks Company, the world's largest manufacturer of muscle-building food products, has developed a breakthrough product, Muscle Juice, which their researchers boast is far better than any other on the market. They know their targeted consumers—professional and amateur bodybuilders—already have a wide variety of products to choose from and are reluctant to embrace new products, particularly those that claim to be "the best." As a result, Muscleworks decides to conduct a series of laboratory experiments on mice to test the muscle-building powers of Muscle Juice against its leading competitors, Bulk Up and Throb. Muscleworks chooses this course of action because it knows that blanket claims that Muscle Juice is "the best product on the market" or that "this product will put a bounce in your step and a smile on your face" are not claims to impress the typical bodybuilding product consumer. Today's hardnosed buyer wants to see studies that say Muscle Juice is "50 percent more powerful than Bulk Up or Throb with twice the gamma radiation of any other brand." Price conscious consumers, for example, are particularly interested in the cost competitiveness of the products they buy. They do not want to read an advertisement that says Muscle Juice will "make you happy" or "help you find eternal peace." Instead they want to see a starburst emblem proclaiming "two for the price of one" or "one dollar off."

Anyway, the scientists at Muscleworks begin their task in earnest by setting up a laboratory where various groups of rats will be force-fed enormous quantities of Muscle Juice, Bulk Up, and Throb to see which mice show the most muscular development in the shortest possible time. Muscleworks wants to be able to trumpet the results of these laboratory experiments, so it will take special pains to ensure that no one can snipe at the way the experiments are conducted. The scientists will group the mice so that each of the different test groups are matched as evenly as possible. This means the Bulk Up and Throb groups will have as many "tough" mice who can chew through telephone cables with one bite as the Muscle Juice group. Of course the Muscleworks team may rethink this strategy of using a level playing field if the results do not show Muscle Juice to be the best product. It may then see the wisdom of putting all of the "sissy" mice on the Bulk Up and Throb teams if only to "stack the deck."

Let us suppose the tests go forward and the three products are given to three groups of mice. The scientists will then track the results at periodic intervals, such as every two weeks. Over time they will observe measurable results as some of the mice bulk up into true "he-rodents." Other mice, however, may not react so favorably to the daily force-feedings and will begin to wear cardigan sweaters and bow ties. Because the results with each individual mouse will vary, the scientists will want to focus on the total increase in bulk recorded by each group. Each mouse would be weighed and any changes in its weight recorded. Similarly, one of the research assistants would tape measure each mouse's length (from nose to tail), torso girth, head circumference, and limbs. All the measurements would be noted so that the changes in each group as a whole could be tabulated.

Once the testing was completed and each mouse had consumed the equivalent of 38,000 gallons of either Muscle Juice, Bulk Up, or Throb, then our results could be plotted on a two-dimensional graph. For our purposes, we could have the vertical axis measure the aggregate change in mouse weight and the horizontal axis measure the passage of time. We could also compare each product by plotting the points in different colored inks or using lines of different textures on the same graph.

Once the results were plotted, we might find that the Muscle Juice was somewhat more effective in causing the mice to gain weight than either Bulk Up or Throb. The Muscle Juice mice might even walk with more of a swagger than either of the other two groups of rodents. We would have to avoid making the assumption that the weight changes were caused by the changes in time simply because they occurred over a given period of time. Similarly, we would have to be careful about attributing the differences in aggregate weight gain to the consumption of the three products unless we could be certain that all other conditions surrounding the test had been held constant.

Any capable scientist will do everything possible to make sure that the test conditions are the same for each experiment. As a result, there should be no other differences that would otherwise muddy the apparent connection between product consumption and aggregate weight gain. We would not expect to find that the water given to the Muscle Juice mice was from a crystal clear mountain stream whereas the water given to the Throb mice, for example, was drawn from the drainage field of a stable. If all three groups of mice are on an even playing field, we can be fairly confident that the only variable is the particular product ingested by the mice.

The other significant issue that will arise is whether the relationship exists in a nonlinear form. In other words, we may find that the aggregate weight gain of each group of mice rises for various periods of time but then declines due to such things as their revulsion at having to consume such huge quantities of bodybuilding products. In such a situation, we might find the lines on the graph for each of the three products would initially rise, peak, and decline. These parabolic-shaped lines are obviously not linear but resemble something approaching semicircles. The lack of a linear relationship does not mean there is no relationship at all between the two variables. Instead it merely means that the relationship between the two variables does not continue to change in the same way, without limit, as would be the case with a positively sloped line showing progressively greater weight gains by the mice.

It seems almost intuitive that we would be unable to sustain such a linear relationship, particularly when dealing with living creatures such as mice. For one thing, there are certain inherent limits on the amount of weight a group of mice can gain no matter how much bodybuilding product we give them. Of course this statement implicitly assumes we are dealing with normal-sized mice and not supermice, which stand 50 feet tall and would think nothing of crushing a house with a swipe of their massive tails. If all we have are regular mice, which can grow no larger than, say, the size of a tape dispenser, then we would be justified in assuming that too much Muscle Juice or Bulk Up will have adverse consequences. The parabolic-shaped lines on a graph indicating the aggregate weight increases for each group of mice over time would suggest any number of possible explanations. It could indicate the mice would develop some type of resistance to the product as they consumed greater and greater quantities so that their initial weight gains would, over time, disappear. This reduction might be caused by the mice's inability to keep down additional amounts of these products or perhaps some internal reaction to these products. It might even be caused by the inability of the mice to digest these products beyond some minimal amounts. In any event, this rise and fall in weight gain would illustrate a nonlinear relationship that would change over time (as shown by the graph) but would not continue to increase without limit.

Regression

What we call regression actually began with the investigations of the English scientist Sir Francis Galton—not the famed psychoanalyst Sigmund Freud. A cousin of Charles Darwin, Galton had wondered why there is a tendency in the general population for the children of taller parents to be shorter than their parents and for the children of shorter parents to be taller than their parents. These ponderings led Galton into some of the earliest studies of what is now known as the science of heredity. The fact that the children of taller (shorter) parents did not continue to grow ever taller (shorter) with each succeeding generation alerted Galton to the notion that there are certain self-corrective elements in humanity's genetic pool that limit the extreme heights to which humans beings will grow. This is why giants are not found in great numbers.

Although the delineation of these factors would have to await the arrival of later generations of medical researchers, Galton was perceptive enough to see that the population did tend toward a band of average heights. This realization that there is a fairly narrow range of heights for humans prompted other fertile minds to consider the ideas that underpin the concept of regression.

Statisticians use regression lines to identify linear relationships between two variables. The so-called regression line can be viewed as a graphical "average" of all of the different plotted points on a graph. However, the use of the word "average" is not really appropriate here, because we are talking about something other than simply adding up all of the vertical and horizontal coordinate numbers on a graph and dividing them by the number of such points. The regression line is something different. It is the line that minimizes the differences between all of the points on a graph. Even though no single point may lie within this line, the regression line represents what statisticians refer to as the line of least squares. It is the line that encompasses the least amount of deviation from each of the plotted points on the graph while still managing to express a linear relationship between the variables.

Perhaps a simpler way to think of the concept of regression is to imagine an array of cones placed in a parking lot. These cones would be the three-dimensional counterparts to the points on our two-dimensional graph. Suppose we have hired a ten-year-old boy to ride the straightest possible line between the cones. The line traced by our junior bicyclist between the obstacles would be the regression line. This is not to say it would be a perfectly straight line because our test pilot might find it necessary to veer to the left and right as he made his way through the cones. Still it would provide us with a pretty good idea of what statisticians are trying to do when they try to draw a regression line that minimizes the distances among the various points on a graph.

Certain important features of the regression line can be expressed using the equation $y = ax + b$. Although we promised not to sneak in menacing-looking equations, most people would agree that this equation is comparatively benign as equations go. This equation basically tells us both the slope of the regression line (variable a) and the point at which the regression line intercepts the vertical (y) axis (variable b). However, this is only part of the labor involved in regression analysis. Statisticians also use a rather forbidding equation beyond the scope of our discussion to calculate the linear correlation coefficient that will reveal the intensity of the relationship between the two variables we are studying. Hence, a greater coefficient will indicate a more intense degree of correlation between the two variables, whereas a smaller coefficient will similarly indicate a less intense degree of correlation. These equations are somewhat cumbersome to explain in this type of book, so we will simply point out that they provide a mathematical expression for quantifying the relationship between the two variables on the graph.

While regression analysis is useful for describing a relationship between two variables, it does have certain limitations, reducing its usefulness. Regression lines are ostensibly geared toward revealing linear relationships between two variables. Thus they are not very helpful when there is no linear relationship between two variables. No linear relationship? Is such a thing possible? Yes! One can certainly have a two-dimensional graph in which the variables on the vertical and horizontal axes do not have a precise linear relationship.

Suppose we have found some funds to pay for an experiment to determine whether there is a relationship between the level of one's income and the amount of Zippy Cola one can consume. Now this study has certain elements that must be considered, including the degree to which Zippy Cola is viewed as an "off-brand" cola product such that wealthier consumers might prefer the more widely advertised brands such as Fizzo and Peppy. Similarly, our researchers would have to take into account the degree to which the consumption of soft drinks in general and cola products in particular decline with increases in income. We might find there is absolutely no relationship between income and the consumption of Zippy Cola. In other words, there might be extremely wealthy people who cannot wait to pour their first bottle of Zippy Cola over their corn flakes in the morning as well as extremely poor people who would not drink a bottle of Zippy Cola even if they were stranded in the middle of the Sahara desert for a week without food and water. There might also be very wealthy people who would sooner spend Christmas with their children than put a bottle of Zippy Cola to their lips or very poor people who would mortgage their hut for a case of Zippy Cola. We might very well find on a two-dimensional graph (with the vertical axis denoting increasing incomes and the horizontal axis delineating increasing numbers of cases of Zippy Cola consumed) that the plotted points (where each point reflected one respondent's statement regarding his or her income and the amounts of Zippy Cola consumed) would be all over the proverbial map. We might have a fairly even distribution of points throughout the graph, suggesting the absence of any general correlation between the incomes of the respondents and the quantities of Zippy Cola they consume.

If the points are scattered all over the graph, then they obviously do not reflect a generally high degree of correlative intensity because they do not fall on or near a single line. (We need to recall that the intensity of the correlation is defined by the closeness with which the points approximate a single line.) The more dispersed the points, the less credible the argument that a linear relationship exists. Assuming that the points are very widely dispersed, it would be impossible to derive any single line that would express a convincing linear relationship between all of the plotted points. It would be a little like trying to draw a single line in a pond to minimize the distances of all of its molecules.

Regression analysis also provides statisticians with a means for making predictions. However, we have to remember to stay within the general parameters of our sample data so that we do not start making absurd predictions that have no basis in fact. If our Zippy Cola diagram showed, for example, that there was a direct relationship between the amount of one's income and the quantities of Zippy Cola one consumed in a given year, then we might be tempted to conclude that this positive relationship between the two variables continued without limit. Suppose the data we had collected by interviewing our respondents showed the following: a person who earned $10,000 per year drank 10 cases of Zippy Cola per year; a person who earned $20,000 per year drank 20 cases per year; and a person who earned $30,000 per year drank 30 cases per year. We might naturally conclude that this relationship would continue upward without limit so that a person who earned $50,000 per year, for example, would consume 50 cases per year. The problem with this analysis is it assumes anyone who has a middle- or upper-class income is essentially hooked up to a intravenous feeding tube, which is, in turn, connected to a 50-gallon drum of Zippy Cola. Obviously, it would be very difficult for any normal individual earning $100,000 per year to drink 100 cases per year (an average of 8 cans of cola per day). We would really separate the serious aficionados of cola drinks from the mere amateurs if we could find an individual who earned $500,000 per year and drank 500 cases per year (an average of about 40 cans per day). No doubt this type of enthusiastic consumption would warm the hearts of even the coldest boardroom warriors at the headquarters of Zippy Cola. Such an extrapolation would be absurd because no one in their right mind would attempt to drink such high quantities of any liquid—including Zippy Cola. Not even Marv Zippy, the son of the company founder, Zebulon Zippy, would try to undertake such an outlandish feat. We would have to be careful to limit our predictions to those parameters in which the relationship between income and consumption is borne out by real world observations. Otherwise, we could end up looking very foolish and finding ourselves the butt of many "dumb statistician" jokes in future issues of *Statistics Today*.

How do these concepts of correlation and regression fit into the general landscape of statistics? As with such things as the standard normal distribution, these concepts are tools to help us organize and interpret data. Statistics deals with probabilities, not certainties, and its value as an analytical tool is limited by the quality of the information being considered. As we have seen, the extent to which a sample can be said to represent the population studied is a primary determinant of whether the inferences we can ultimately draw about the population will be worth the paper on which they are written. Similarly, correlation and regression analysis play important roles in the ordering of data into comprehensible and useable forms. They help to identify the strength or weakness of the relationship between the variables examined, and thus provide at least some insight into the relationship between those variables, if not the actual causal connection—if any—between them.

Epilogue

Conquering Statistics?

W e have spent much of this book talking about the central concepts of statistics and the ways in which commerce and industry have benefited from the use of these tools. At the same time we have tried to have a little fun along the way while still showing appropriate respect for the subject matter. It is my hope we have been at least moderately successful in demonstrating both the utility and beauty of statistics.

This is not to say this book has covered every aspect of the field of statistics because our aim has been considerably more modest. We did not want to write a textbook on statistics because that would have defeated our purpose of narrating a lighthearted tour through the world of statistics. We were not concerned with delving into every facet of statistics but merely with touching upon its most general and central concepts. The level of the presentation has been very basic as have been the illustrations of the concepts. Our sporadic attempts to liven the book with humor have been made not to disparage the subject matter but to provide a somewhat more accessible account of the subject.

Statistics is a sort of mathematical map that enables us to impose order on the disorganized cacophony of the real world. It provides us with powerful analytical tools for drawing conclusions about the mass of data that permeate our modern advanced society. As we have seen in this book, statistics is critically important when solving problems relating to quality control in manufacturing. Statistics also makes it possible for us to take small samples and make accurate generalizations about their respective populations as a whole. Of course we need to recall that statistics is ultimately grounded in probabilities. It is not concerned with certainties but *near* certainties. We used the degree of confidence to control the closeness with which we wish to approximate 100 percent certainty. We did not seek such perfection because it comes only with taking into account the features of the entire population from which our samples are to be drawn.

Why has statistics received such poor press if it has been so beneficial to society as a whole? One possibility is that its mathematical nature requires a certain degree of discipline in order to master its intricacies. Needless to say, mathematics in general has suffered from the same bad press because it is exacting and, hence, unforgiving of ambiguity and error. Unlike an essay discussing the merits of a book such as *Lady Chatterly's Lover*, in which a wide spectrum of differing and even contradictory opinions could be entertained in a credible manner, the test for solving a mathematical equation is much more blunt. One is either successful or unsuccessful in solving the equation; no credit is given for a "sporting try."

Statistics also suffers in the public's eyes because it is a rigorous subject and requires its students to follow certain very specific procedures to obtain the solution to a problem. This predictability and invariability should be a source of comfort. Instead, it appears to intimidate many people who lack the desire to learn the operations that would permit mastery of the subject.

Statistics is also seen as being cold and sterile. Perhaps this criticism is based upon the idea that it, like all mathematics, is based upon a finite number of distinct mathematical principles. Because these principles are immutable, all that flows from this theoretical foundation is presumed to be dusty and dreary, as though it were the antithesis of color and light. Of course, one person's idea of cold and sterile is another person's idea of reassuring and familiar, so this perception is probably not shared by all individuals. In some sense, then, it harkens back to the idea of statistics as being the province of those who cannot be creative or dream in color.

Statistics is a tool that helps us to learn about the world around us. It can be used in as creative a manner as we desire to solve complex problems faced by business and industry. It can also be marshalled to gauge the shifting sentiments of public opinion to predict the outcome of elections. The idea that it is somehow too pristine or at least not part of the "real world" is patently absurd, as shown by the myriad of ways in which it can be brought to bear to yield answers to the problems we encounter in our daily lives.

The title of this book suggests it is a vehicle for subjugating statistics. This is not the intention because our goal is to wander through some of its most fertile fields and learn about its most fundamental concepts. One who reads this book and no other will not be able to obtain a position as a college professor of statistics. We are hopeful it will provide the novice with a general grasp of the subject's reach and enable that person to apply its principles to the daily events of his or her life. If anything, we hope the review of this subject will encourage persons to think more critically about everything from surveys to news reports to advertisers' claims. After all, we are all bombarded with an incredible variety of information every day. Some of it is useful, some of it forgettable. We are not always able to determine easily whether the information is credible. Perhaps the greatest single benefit of studying statistics is that it forces us to question the basis for the conclusions supposedly yielded by scientific studies or laboratory tests or public opinion polls. It thus encourages us to move from the question of whether we agree with the conclusion to the more important question of whether we think the conclusion was derived in a credible and defensible manner.

The writing of this book also served as a reminder of the ease in which the general ideas about statistics can be understood and put to profitable use. It would be difficult to think of a branch of the sciences that has provided greater benefit to the industrial world. Certainly statistics can claim to have as great a stake in the modern economy as chemistry and physics because statistics alone makes possible the functioning of the market economy and the management of massive multinational corporations.

Anarchists may bristle at the thought that statistics is propping up the entire global economy. But this statement is not very far from the truth. Nearly every industry and form of commerce requires at least some type of statistical analysis. Statistics makes it possible to manipulate and utilize in a meaningful way massive amounts of information. It also helps to provide guideposts for organizing this information in ways that may lead to additional insights and solutions that would not have otherwise been possible without these techniques.

Sir Isaac Newton reportedly compared his own unparalleled accomplishments in the sciences (which included his theory of gravitation, his laws of motion, and the discovery of the calculus) as being nothing more than akin to finding an occasional pebble on the beach while the great ocean of truth lay all around. Statistics is a sort of intellectual probe that makes it possible for us to gather information about the samples we find scattered along the way so that we can make educated guesses about the nature of the populations from which they are drawn.

Statistics cannot claim the exactness of physics and pure mathematics because it is, to some extent, a social science. Certainly its pedigree can be found in mathematics, but its subject matter is so often the seemingly capricious and unpredictable world of human beings. As a result, it does not quite stand shoulder to shoulder with physics and mathematics in terms of prestige. After all, physics is the science of the smallest bits of matter as well as the largest galactic clusters. Mathematics is the mistress of physics, providing the language by which most of the fundamental processes in nature can be explained in a clear, quantifiable manner. Statistics does not fit in neatly with these two sciences, despite its mathematical roots. It can be used to analyze a wide variety of physical processes even though the focus of our book has been on human activities. As statistics ventures beyond the certainties of physical phenomena to grapple with the vagaries of human behavior, it somehow seems to be less fundamental and essentially less important than its more famous parents.

Where statistics ranks in the pantheon of sciences is not particularly important because it does play such an important role as an applied science. In many ways, it has a more immediate day-to-day impact in our lives than does any other science. Certainly we would be upset if the laws of physics no longer continued to apply so that the sun began to burn out. As long as the sun continues to rise each morning, we have the luxury to wonder about the representativeness of samples relative to their parent populations and engage in carefree banter about confidence intervals and inductive reasoning. If the universe continues to run on as it has for billions of years, then we will have the leisure time to continue using our statistics to gather samples for drawing inferences about any kind of population we choose.

As far as the coverage of this book is concerned, I would be the last person to suggest we have delved into every crevice and cranny of the field of statistics. My purpose in writing the book was to provide a broad brushstroke of the subject and not to go into excessive mathematical detail. It was my desire that the book be both accessible to a general audience and sufficiently detailed so that any reader could at least walk away with a certain knowledge of the guts of statistics.

This approach also necessitated that a number of subjects such as contingency tables, nonparametric statistics, and multiple regression be omitted from discussion. My desire that the book be as devoid as possible of mathematical equations also prompted my omission of a number of equations that are critically important to the work of statisticians.

The attempt to season the book with stabs at humor was something of a risk because statistics is not usually thought of as being a jovial enterprise. It seemed silly to try to present the subject in a dry, textbook-like fashion because it is precisely this approach that has given statistics its poor public image. The utilization of many fanciful examples was designed to encourage the reader to have a little bit of fun while following the narrative but also to bear in mind that the subject is only as dull as one cares to make it. If one approaches it with an open mind and an upbeat perspective then the imposing edifice that once appeared to tower upward to forbidding heights will become more accessible and understandable.

So it is with a fond farewell that we bring this affectionate survey of statistics to an end. It has been more than a year since my editor and I decided over a lunch in New York that writing a lighthearted guide to statistics would be a worthwhile book project. Of course it took some time to figure out how best to structure this book and the presentation of the subject matter. Certainly I did not want the tongue-in-cheek approach that I so dearly wished to use to be misconstrued as a clumsy attempt to be cute or clever. I am quite aware that one person's humor is another person's aggravation. Moreover, humor is very subjective and can be misinterpreted, particularly in an age in which people seem to take offense more readily at less and less. At the same time, I have always been concerned about the ease with which the general public can learn about the sciences and it seemed that such a whimsical approach was justified.

As with every other book I have been involved with—either as an author or co-author—I have benefited from the works of many men and women who have made possible the magnificent pantheon of concepts and ideas we call science. By the same token, I alone am responsible for the errors or omissions contained within this book. Whether this book is a worthy product and whether it succeeds as what is hopefully an elegant but unpretentious narrative that explores a subject truly deserving of the respect of the general public is a judgment that must be left to the reader. For those who wish to delve deeper into the mysteries of statistics, however, I would recommend at least a cursory review of any of the titles contained in the bibliography.

Endnotes

Chapter 1

1. John Graunt, Preface to *Natural and Political Observations Mentioned in a following Index, and Made upon the Bills of Mortality*. London, 1662.
2. John Graunt, Letter to John Lord Roberts in *Natural and Political Observations*, supra.
3. Lowell J. Reed in the *Introduction to Degrees of Mortality of Mankind* by Edmund Halley. Baltimore: Johns Hopkins Press, 1942, p. iv.

Chapter 4

1. Tom Ainslie, *How to Gamble in a Casino.* New York: Fireside, 1979, p. 26.
2. Ibid, pp. 26–27.

Chapter 5

1. Mario F. Triola, *Elementary Statistics,* 5th ed. Reading, MA: Addison-Wesley, 1992, pp. 187–188.

2. Jerry L. Patterson, *Casino Gambling.* New York: Coward-McCann, Inc., 1982, p. 33.
3. Ibid, p. 34.
4. Ibid.

Chapter 7

1. See Mario F. Triola, *Elementary Statistics*, 5th ed. Reading, MA: Addison-Wesley, 1992, pp. 289–290.

Bibliography

Ainslie, Tom, *How to Gamble in a Casino*. New York: Fireside, 1979.

Ball, W. W. Rouse, *A Short Account of the History of Mathematics*. New York: Dover, 1960.

Barstow, Frank, *Beat the Casino*. New York: Pocket Books, 1986.

Bell, Eric Temple, *Men of Mathematics*. New York: Simon & Schuster, 1937.

Boyer, Carl B., *A History of Mathematics*. Princeton, NJ: Princeton University Press, 1985.

Byrkit, D., *Statistics Today: A Comprehensive Introduction*, Menlo Park, CA: Benjamin/Cummings, 1987.

Campbell, S., *Flaws and Fallacies in Statistical Thinking*. Englewood Cliffs, NJ: Prentice-Hall, 1974.

Freedman, D., R. Pisani, and R. Purves, *Statistics*. New York: Norton, 1978.

Freund, J., *Modern Elementary Statistics*, 7th ed. Englewood Cliffs, NJ: Prentice-Hall, 1988.

Huff, D., *How to Lie with Statistics*. New York: Norton, 1954.

Kasner, Edward, and James R. Newman, *Mathematics and the Imagination*. New York: Simon & Schuster, 1940.

Kimble, G., *How to Use (and Misuse) Statistics*. Englewood Cliffs, NJ: Prentice-Hall, 1978.

Kline, Morris, *Mathematics and the Physical World*. New York: Dover, 1959.

Motz, Lloyd, and Jefferson Hane Weaver, *Conquering Mathematics*. New York: Plenum and Avon, 1991.

Motz, Lloyd, and Jefferson Hane Weaver, *The Story of Mathematics*. New York: Plenum and Avon, 1993.

Newman, James R., *The World of Mathematics*. New York: Simon & Schuster, 1956.

Patterson, Jerry L., *Casino Gambling*. New York: Coward-McCann, 1982.

Paulos, J., *Innumeracy: Mathematical Illiteracy and its Consequences*. New York: Hill and Wang, 1988.

Randall, John Herman, Jr., *The Making of the Modern Mind*. New York: Columbia University Press, 1976.

Singh, Jagjit, *Great Ideas of Modern Mathematics: Their Nature and Use*. New York: Dover, 1959.

Smith, D.E., *History of Mathematics*. New York: Dover, 1951.

Strunk, Dirk, *A Concise History of Mathematics*. New York: Dover, 1987.

Triola, Mario F., *Elementary Statistics*, 5th ed. Reading, MA: Addison-Wesley, 1992.

Index